Moving UP North

Weronika Anna Kusek

Moving UP North

Immigrant Experiences of Professionals in the Upper Peninsula of Michigan

Springer

Weronika Anna Kusek
Earth, Environmental, and Geographical Sciences Department
Northern Michigan University
Marquette, MI, USA

ISBN 978-3-031-95700-0 ISBN 978-3-031-95701-7 (eBook)
https://doi.org/10.1007/978-3-031-95701-7

This Springer imprint is published by the registered company Springer Nature Switzerland AG
The registered company address is: Gewerbestrasse 11, 6330 Cham, Switzerland

For my children, Eddie and Ania, and for everyone who grew up with immigrant parents.
This book is to inspire you to be courageous, resilient, and persistent wherever life takes you.
Contribute to your community, respect people around you, build a strong network of friends, and enjoy all the beautiful places.

Acknowledgements

This book is a result of many interactions with people who live and work in Marquette, Michigan. When I first visited Marquette, MI, in the winter of 2013, like many immigrants, I possessed very little knowledge about Marquette or the broader "UP." As a geographer, I was aware that the Upper Peninsula of Michigan existed on the map, but I did not know much about the history or reality of everyday life in this region. It took me a few years to learn about the history of Marquette and the surrounding area, the people who influenced it, and how to take advantage of the natural beauty of the region to make my life in my new hometown truly enjoyable. Over the last 10 years, I met many people who not only introduced me to the Upper Peninsula but rather taught me "the Yooper way" to enjoy all that this place has to offer.

I also learned the history of Marquette by reading the vast work about the Upper Peninsula authored by Professor Russell M. Magnaghi. Much of my knowledge is also a result of support from colleagues from Northern Michigan University who offered countless lectures during a variety of symposia and conferences.

Institutions such as the Beaumier Upper Peninsula Heritage Center—a historical museum and educational facility on the campus of Northern Michigan University—and the Marquette Regional History Center have been crucial to building my understanding of Marquette and the Upper Peninsula region.

This book would not have been possible without all the wonderful people who agreed to be interviewed for this book project. I am thankful for their trust in my work, for sharing their most personal stories, and for the many hours they sacrificed to provide me with the material needed to write this book.

I would also like to recognize the artistic contribution of my friend Lali Khalid who is a visual artist at the School of the Art Institute of Chicago. Lali offered to collaborate on my book project by sharing her photography as a way to provide readers with a glimpse into the stunning environment of the Upper Peninsula of Michigan. I truly admire Lali's talent to capture the natural beauty of the region.

My ability to understand my interviewees and relate to their passion for the Upper Peninsula would have not been possible without many of my own friends from Marquette who helped me along my own Upper Peninsula journey when I enthusiastically accepted a job at Northern Michigan University. Jessica and Jonny invited me to spend time at their camp on Lake Independence so I could enjoy firepit-side conversations at sundown. They also provided me with the opportunity to camp in a tent on their property and be awe-struck by sights of a nighttime UP sky unobstructed by light pollution characteristic of a big-city sky. Maureen introduced me to the culture of boating on an inland lake near Gwin, MI. Michele took me on a trail run to expose me to the running culture of Marquette. I also spent hours on a boat with Amadeo and Mary who were teaching me how to enjoy the "big lake" and proved that under the right wind conditions the water of Lake Superior is not only crystal clear but also almost warm. Kara and Marcin showed me the beautiful Garden Peninsula and left me in awe of the spectacle of Northern Lights. Susy and Dick's love for skiing was contiguous and encouraged me to return to skiing after years of hiatus from winter sports. I have also leveraged my wonderful administrative assistant Jana as my Yooper expert who provided me with countless advice on where to find a hairstylist, where to get winter tires, and where to buy the best pasties.

Finally, I would like to thank my husband Jacek who encouraged me to take an academic position at Northern Michigan University. He supported me emotionally throughout my academic journey. And, as always, while I was completing this book project, he was patiently listening to all my ideas for this book, reacting to all my stories and reflections that I shared with him while I was completing the interviews, and he was the calming force for my anxiety when I was questioning the timeframe and whether this project would even be possible.

Contents

1 **Introduction** 1
2 **Zosia** 15
3 **Hema** 23
4 **Ritesh** 29
5 **Claudia** 35
6 **Gana** 41
7 **Miguel** 47
8 **Yan** 53
9 **Wael** 59
10 **Zhenia** 65
11 **Sonia** 71
12 **Conclusions** 79
Index 85

About the Author

Weronika Kusek is an associate professor of geography in the Earth, Environmental, and Geographical Sciences Department at Northern Michigan University. She is a native of Poland and an immigrant to the United States, and her academic work focuses on the individual experiences of immigrants. Weronika is married and has two children who identify as "Polish Yoopers."

Chapter 1
Introduction

Introduction

Transnational migrants transcend national boundaries in search of better educational, professional, and personal opportunities and experiences. While today there are various regions in the world that are developing as migrant destinations, the United States remains a country with the highest number of immigrants with 51 million people, both documented and undocumented immigrants (Camarota & Zeigler, 2024). Although the change in policy, immigration enforcement, and public rhetoric under the second Trump administration have undermined the confidence of immigrants and visitors about arriving to or remaining in the United States, the United States still remains a country with the highest immigrant population. The historic volume of migration to the United States means that migrants represent various backgrounds and reasons for migration and pursue highly varying paths for establishing life in this country. That said, most current research and media coverage focus on chain migration among laborers and undocumented immigrants in urban settings, leaving the nuanced challenges of professional immigrants in rural locales largely unexplored.

There has also been vast literature that focuses on socio-economic polarization in the context of global elites who influence urban areas (Sassen, 1991; Hamnet 1994; Ray, 2003). But little attention has been given to professional immigrants who settle in small towns and rural areas of the United States, especially in the rural North. Recent work focuses on new immigrants who have settled in rural and suburban areas within the South but there is not a substantial piece of research that features new immigrants and their experiences in the northern parts of the United States. Helen Marrow in her "New Destination Dreaming" discusses how the rural context impacts the immigrant experience and how rapid Hispanic immigration influences southern race relations.

W. A. Kusek, *Moving UP North*, https://doi.org/10.1007/978-3-031-95701-7_1

"Moving UP North—Immigrant Experiences of Professionals in the Upper Peninsula of Michigan" attempts to address this gap in literature by focusing on individual immigrant experiences of professionals in the American North and more specifically the Upper Peninsula of Michigan.

Professional immigrants move to a new country with specific skills and specialized training, possess formal advanced degrees, and offer expertise often in fields such as technology, engineering, medicine, academia, or other specialized areas. This type of migration is typically driven by opportunities for career advancement, research, or entrepreneurial ventures, and these immigrants are admitted to the United States under formally legal immigration permits or visitor permits that often provide a path to permanent residency. Professional immigrants often fill critical gaps in the labor market as their legal status in the United States often depends on their ability to prove the need for their skills. They drive technological and scientific advancements and are recognized for contributing to higher industrial productivity. Little is known, however, about their personal experiences in their new hometowns.

Theoretical Framework

Mobility of Professional Immigrants in Geography

Human geographers study how humans interact with spaces they occupy, how they change natural landscapes, and how they culturally transform the communities they live in (Malinowski & Kaplan, 2012). Mobility refers to the ability to move and often involves migratory movements that include push and pull factors or in other words motives to leave someone's home and relocate to a new place (Cresswell, 2011; Kusek & Wise, 2020). Mobility in that sense is often based on privilege as it involves access to education, economic position, and ability to make a choice for oneself. Mobility and migratory movements of humans are at the center of discussions on how human involvement affects spaces, places, and landscapes, how sense of places is being created and recreated, and how spaces become places through human involvement and human experience (Wright, 1947; Tuan, 1974, 1977; Kusek & Wise, 2020). A sense of place refers to the emotional and subjective connections people have to a particular location. Sense of place can be positive, when an individual feels safe, accepted, and comfortable, but it can also be negative when one experiences discrimination or placelessness. It is about the emotional and psychological connections people develop with a place, shaped by its physical characteristics, history, culture, and most importantly individual experiences. The term sense of place is used here to set the stage for describing how immigrants perceive and experience a small town in a rural locale. The main focus of this book is to explore the stories of current professional immigrants in Marquette, MI, their experiences, and how they make their home away from home. Before delving into individual immigrant experiences, it is important to define the concept of *place* from a human geography perspective.

Sense of Place

In the common understanding of the concept, a place is often seen as an exact geographic area defined by coordinates and physical boundaries. That said, for human geographers and other social scientists, *place* has social and cultural significance in the context in which it exists. Cresswell (2011) argues that *place* is not just a physical location but a complex concept that includes both the material and symbolic aspects of space. Human geographers focus their attention on the many activities and cultural transitions that occur in a specific place. They also study social relationships that are created there. For example, physical geographers will define a beach as "a landform alongside a body of water which consists of loose particles such as sand or gravel" but human geographers will focus their study of the beach on how this environment provides opportunities for people to gather, relax, and create memories. They consider the emotional attachment and personal connections that people have to a place.

Place as home has been studied by many scholars for years (Tuan 1977). Bachelard (1994) believes that home determines how we understand other places. In a chapter I wrote in 2016 (Kusek, 2016) I explored how my hometown of Poznan, Poland, predisposed me for my migratory journey. I concluded that as *"I transition locations, place for me combines the features and factors of the external with the ideas and activities that I bring with me. Place is something I experience, but also something I recreate and practice. This interaction between the internal and external allows me to believe that cumulatively and chronologically, places have shaped who I am"* (Kusek, 2016, p. 70).

Furthermore, Cresswell (2011) argues that "place" is not just a neutral space but it is a dynamic, complex, and socially constructed concept. Place is shaped by both human interactions and the physical environment, and it has emotional and cultural significance. In this book, the professional immigrants discuss the places that have shaped their personal and professional identities and they emphasize the importance of place while choosing to continue their immigration journey and settling in Marquette, MI. They share their attachment to the Upper Peninsula and the appreciation for the beauty of natural landscapes while in many cases creating more permanent roots in a new community and applying their personal touch to their daily environments.

Marquette, MI

Marquette County is the biggest economic hub in the Upper Peninsula of Michigan (colloquially referred to as the "UP") and for decades it has been attracting people from many corners of the United States and the world. Marquette is a major port city in Michigan thanks to its access to Lake Superior and has been instrumental to the iron ore industry and the Great Lakes shipping business for over 100 years (Magnaghi, 2017) *(please see* Image 1.1*)*.

Marquette is a town of incredible nature offering a variety of outdoor activities: hiking, mountain biking, hunting, cross country and alpine skiing, and snowmobiling, as well as a variety of summer sports including kayaking, boating, golfing, surfing, or simply enjoying time on many of the beaches in the city and region (Travel Marquette, 2024). All these activities are complemented by a robust cultural and entertainment scene in the city of Marquette and the campus of Northern Michigan University that together offer a variety of cultural and sporting events; community, music, and food festivals; nationally recognized dog sledding races; farmers' markets; farm to table restaurants; local breweries; and much more. The variety and quality activities attract not only tourists but also many students and professionals who embrace city activity options balanced by the opportunity to engage with the outdoors.

Marquette is a city of 21,000 people located in one of the most remote and rural parts of the United States. Despite its remote and rural location, Marquette is well recognized as a tourist destination and, as the latest statistics indicate, an

Image 1.1 Ore Dock in Marquette, MI. (Photography by Lali Khalid)

increasingly attractive destination for international migrants. 2024 numbers indicate that Marquette experienced an increase in its international population (Michigan League for Public Policy, 2024). Marquette County has a diverse immigrant population, with people coming from different regions of the world. In contrast to the more pronounced chain migration in other parts of the country, immigrants to Marquette represent a variety of geographic regions: Asia, Sub-Saharan Africa, Europe, and South America. According to the Michigan League for Public Policy, "48% of immigrants in Marquette County arrived in the United States before 2000, and since 2010, the number of immigrants has increased by 18%" (Michigan League for Public Policy, 2024). In 2020, Marquette County had 1435 immigrants of whom 873 were naturalized citizens and 562 were noncitizens that included visa holders, lawful permanent residents, refugees and asylees, and undocumented residents as well.

Part of what opens Marquette to international immigrants is the presence of large and significant employers. Marquette is home to Northern Michigan University, the UP Health System which operates a level 2 trauma hospital and serves as a regional medical center, and the Eagle Mine, all of which hire international professionals.

Professional immigrants also live in other parts of the Upper Peninsula with a significant concentration in the Houghton area where the Michigan Technological University is located.

The goal of this book is to focus on some of those individuals and spotlight personal stories and the diverse migratory journeys of immigrant professionals who live and work in Marquette. The objective is to highlight how these individuals play a critical role in supporting the functioning and growth of this community, but also to identify the unifying and shared experiences of people who come from different ethnic backgrounds, cultural contexts, and geographies but equally made Marquette and the "UP" their home. Among professors, physicians, and engineers, there are also foreign spouses of local residents who tell their stories to provide a unique perspective on assimilation, cultural adjustment, homemaking, sense of belonging, and longing for home.

History of Immigration in the United States

To fully understand the significance of stories of immigration to Marquette featured in this work, it is important to place local immigration patterns within the broader context of US immigration history. Many argue that today the United States is facing the biggest anti-immigrant crisis since the 1920s which will likely have an impact on professional immigrants to the United States although that outcome is currently undetermined. Despite the onset of a new sentiment towards immigrants in the United States, it is unclear if professional immigrants will choose to immigrate to countries where more favorable attitudes are being expressed, or if the United States will still remain the main immigration hub for those who possess professional skills, talents, and educational attainment.

It is important to remember that the history of immigration policy in the United States is complex and it has evolved over the last few hundreds of years, not always in a linear fashion. Historic changes in immigration policy are a direct reflection of the changing economic situation, the needs of the US job market, political climate, and attitudes that Americans have towards the cultural, racial, and religious backgrounds that immigrants come from.

In the Colonial Era immigration was largely unregulated. The first immigrants arrived from Europe and were mainly from Great Britain with a few arriving from Spain, the Netherlands, and France (Baxter & Nowrasteh, 2021). At that time, Africans were brought involuntarily to the colonies as slaves. The first federal law, the Naturalization Act of 1790, regulated citizenship by allowing "free white persons" to become US citizens (Baxter & Nowrasteh, 2021; Harzing et al., 2009; Hass et al., 2020).

In the nineteenth century, immigration increased significantly attracting Catholics from Ireland and Germany. This immigration influx was driven by economic opportunity in the United States as much as by famine in Ireland and political unrest across the European continent (Baxter & Nowrasteh, 2021; Harzing et al., 2009; Hass et al., 2020).

In 1882, the US Congress passed the Chinese Exclusion Act which is considered to be the first federal law to clearly restrict immigration based on race and nationality. This act targeted Chinese immigrants who had been arriving in large numbers in the western United States. The act banned Chinese laborers from entering the United States and was renewed several times before being repealed in 1943 (Baxter & Nowrasteh, 2021; Hass et al., 2020).

Ellis Island in New York became the largest immigration entry point in the United States between 1892 and 1954 and it allowed millions of immigrants, mostly from Europe, to enter the United States (Baxter & Nowrasteh, 2021). The Immigration Act of 1924 was the first attempt to limit immigration to the United States on a large scale. This act set strict quotas based on national origins. It favored immigrants from Western and Northern Europe while restricting those from other parts of the world (Baxter & Nowrasteh, 2021). Immigration dropped during the Great Depression as jobs in the United States became scarce, and many immigrants returned to their home countries. Later, during WWII, in 1942, the Bracero Program was established to allow Mexican workers enter the United States to fill seasonal labor shortages in the agriculture sector (Baxter & Nowrasteh, 2021, Harzing et al., 2009; Hass et al., 2020).

The Post-WWII Era brought several changes to the US immigration policy. The Immigration and Nationality Act of 1952 (McCarran-Walter Act) introduced a more structured and standardized process for immigration. It retained the national origins quota system but also created preferences for skilled workers (Baxter & Nowrasteh, 2021). In 1965, the Immigration and Nationality Act (Hart-Celler Act) focused mainly on family reunification and skilled labor preferences. This act increased the number of immigrants from Asia, Africa, and Latin America (Baxter & Nowrasteh, 2021).

The 2000s brought immigrants from all over the world—South Africa, South Asia, and the African continent—and the United States was often accused of causing brain drain attracting the most talented to serve the US job market and advancing the high-tech sector (Harzing et al., 2009; Hass et al., 2020) yet the popular media focused its attention on those who filled the gaps in the agricultural, construction, and hospitality sectors. With the diverse populations representing a variety of cultural and religious backgrounds in the United States, the services and unequal distribution of laborers across the United States caused less favorable attitudes towards immigrants. Yet it is an undeniable fact that many areas of the United States, and rural areas in particular, heavily rely on professional immigrants to fill the gaps in the job market. Physicians, engineers, professors, religious servants, and many others serve the regions of the United States where American professionals are less likely to choose to locate.

I am finalizing this manuscript in 2025 as the United States is experiencing a significant shift in immigration policy and enforcement. It is likely that a new chapter in American immigration history is being written currently. For some of the people featured in this book, the impact is direct and significant; for others it is at least currently less perceptible in their day-to-day reality. That said, a lot of the events, experiences, and emotions collected in this book are a point-in-time capsule of local and interpersonal interactions that remain valid and set in the past. I would definitely like to follow up on these stories in the near future to present them in an updated context.

History of Immigration into Marquette, MI

The history of immigration to the Upper Peninsula of Michigan is a wide-ranging story, shaped by its natural resources, economic opportunities, and unique geographic location. Before European settlers arrived, the Upper Peninsula was inhabited by various Native American tribes, including the Ojibwe (Chippewa), Ottawa, and Potawatomi (Magnaghi, 2017).

By the mid-1800s the UP experienced a new immigration influx driven largely by the discovery of iron ore and copper, which spurred mining operations. The influx of immigrants was primarily motivated by economic opportunities in these growing industries. Many Cornish and Welsh immigrants arrived to work in the copper mines, especially in the Keweenaw Peninsula. They brought with them mining expertise and food culture (Magnaghi, 2017).

Another immigration wave included Germans who were often involved in agriculture and Irish immigrants who worked in the mines and railroads. Another significant immigrant group included the Scandinavians: Swedish, Finnish, and Norwegian immigrants also settled in the UP, contributing to the region's cultural landscape. The Finns were particularly prominent in the mining industry, and their presence remains influential in local culture (Magnaghi, 2017).

Stories of immigration to the UP often focus on the influx of immigrants from Finland and many Yoopers (colloquial term for people from the "UP") are proud of their Finish origin. This can be seen in the culture of Marquette. Food, homes, and activities resemble those of Nordic European culture. Visitors may find it surprising to learn that in Marquette many historic homes have saunas. Tourists find it interesting to shop at a local Finnish store that sells Nordic home goods, foods, and other items such as sauna accessories, or traditional pastries and pies with locally made fruit jam.

Other groups of immigrants included those who arrived in the early twentieth century drawn by the lumber industries and included a mix of immigrants from Eastern Europe such as Poles and Slovaks who came seeking better economic prospects and political freedom. After World War II the region experienced a decline of the mining and lumber industries which resulted in a decrease of new immigration waves, and in fact, many UP residents left the region in search of better job opportunities elsewhere in the country (Magnaghi, 2017).

Recent Immigration into Marquette, MI

Today, the Upper Peninsula's population is relatively stable, with new immigrants arriving more sporadically compared to the waves of the nineteenth and early twentieth centuries but the new immigrant group is diverse and includes immigrant professionals from all over the world (Michigan League for Public Policy, 2024). They fill critical gaps in the local labor market. Among professors, physicians, and engineers, international migrants also include foreign spouses of local residents. The influx of international professionals to Marquette is not new but it has intensified in the last decade. Prof. Russell Magnaghi published several pieces that described stories of international professionals in Marquette. Among the many people he featured were medical professionals and academics who contributed to the local community over the last few decades.

Dr. Cintamaneni Bhujanga Rao, a physician, immigrated to the Marquette area from India. He described his experience in the UP, the pull factors for migration, and how he enjoyed the calmer, more relaxed lifestyle in the UP as well as the beauty of the surrounding environment, and he emphasized the opportunity that the United States provides to the medical professional community (Magnaghi, 2017). Another professional immigrant, Dr. Nancy Rodgers-Chenoweth, a physician from Canada, installed one of the first X-ray machines in the UP (Magnaghi, 2017).

On May 4, 2024, at the NMU commencement ceremony, the commencement speaker Milton Chen mentioned his father moved to the Marquette area in the early 1950s to work as an engineer for Cleveland Cliffs. He joined another engineer from China and the two families may have been the only Chinese families living in the area at that time according to Russell Magnaghi.

The twentieth-century Marquette attracted immigrant professionals from different parts of the globe. Dr. Miodrag Bata Georgevich, first head of the Department of

Political Science at Northern Michigan University, immigrated to the UP from Yugoslavia in the 1950s. He became a naturalized citizen in 1955 and continued his education. He became a professor of international relations in 1958 at NMU where he introduced the first foreign study programs for international relations majors. He also founded both campus Young Republicans and Young Democrats clubs and was the school's first coach of the alpine ski team (Obituary, 2004).

These individuals are just a few examples of professional immigrants who contributed to Marquette.

Similar Work

This book takes inspiration from the work of Prof. Russell M. Magnaghi and his historic lens to immigration into the UP. I also rely on the work of Alan Marcus (2016) and his book "Transnational Geographers in the United States: Navigating Autobiogeographies in a Global Age" whose geographic perspective informs much of my approach. Marcus focused solely on international academics who live and work in the United States. His work does not address the challenges and opportunities of living in small towns and remote locations. Marcus' book includes a collection of essays as "a way to map personal trajectories and experiences which examine the concept of place at the micro-level."

"Moving UP North—Immigrant Experiences of Professionals in the Upper Peninsula of Michigan" is a monograph based on interviews conducted by the author and offers descriptive and theoretically informed insight into the lives of these international professionals who live and work in Marquette, MI.

This book addresses the concepts of migration, displacement, sense of place, sense of belonging, and homemaking, and the focus is given to professionals who specifically choose to settle in a small town in the Upper Peninsula, a remote place which is often avoided by diverse populations.

This work emphasizes the diversity of countries that the new immigrants who settle in the North arrive from, rather than the focus on a specific population as it is often the case in studies describing the South.

"Moving UP North—Immigrant Experiences of Professionals in the Upper Peninsula of Michigan" is composed of 12 chapters: in addition to introductory and summary chapters, ten chapters focus on individual experiences of international professionals. The ten chapters tell a story of transnational immigrants who live and work in the Upper Peninsula of Michigan. Each chapter is based on an interview with an international professional who lives in the region: some of the immigrants have been in the area for a couple of decades and others are newcomers.

International professionals answer geographical questions such as: How do you define a place where you live and work? How do you make home in a new place? How do you create and recreate a sense of belonging in a remote place where you may be the only representatives of your ethnic/cultural/racial group? How does educational attainment influence the capability of isolated migrants to create a sense of home? How does the lack of other nationals influence the migrants' sense of obligation to represent their culture and country?

Methodology

This book is a collection of oral histories. The oral history methodology relies on the process of collecting historical information through recorded interviews.

All participants featured in this book voluntarily participated in this project. The professional immigrants featured in this book are individuals who have first-hand experienced the process of immigration and established their professional and personal lives in Marquette, MI. To become US residents or citizens, they completed one of the formally available immigration paths set forth by the US Citizenship and Immigration Services Agency of the US Department of Homeland Security.

These professional immigrants represent a variety of geographic regions, diverse professional identities, and diverse cultural, linguistic, and religious backgrounds. They were selected for this project by snowball sampling which is a research technique that uses existing study participants to identify other potential participants. There were general attributes used to select this group: (Bachelard, 1994) must be a documented immigrant, (Baxter & Nowrasteh, 2021) must currently hold a professional appointment, (C.B. Rao Oral History Transcript, 1997) interviewees must represent a balance in perspectives from a gender identity standpoint, (Camarota & Zeigler, 2024) participants will ideally represent as many various global regions as possible, and (Cresswell, 2011) participants must have lived in the Marquette area for at least a year. The participant selection process was aided by the nature of my life in Marquette. This is a small community, it is relatively easy to get access to others in the community through shared connections, personal recommendations of friends, and opportunities for interaction facilitated by the university environment. Overall I approached approximately 25 individuals to ask about their willingness to participate in this project.

All participants of this project were asked open-ended questions and were informed that they could stop the interview at any time. The personal perspective allows to capture individual experiences and narratives often not found in traditional big data research projects on immigration.

It is important to mention that international professionals featured in this book have chosen to live, work, and contribute to the Marquette community. It is an educated group of professionals who have the choice and resources to choose the location where they live. They fell in love either immediately or over time with the Upper Peninsula. Naturally, the professional immigrants featured in this book represent a positive attitude to the place, people, and culture of Marquette. Those who did not enjoy the long winters, who struggled with the remoteness of the place, who were unable to continue their professional careers, or who did not feel comfortable in a community that does not have ethnic neighborhoods have left the UP for professional opportunities elsewhere in the United States or in the world.

To assist the readers with envisioning the natural beauty of the Upper Peninsula of Michigan and to present the places that many participants of this project have directly or indirectly referenced in their stories, I collaborated with visual artist Lali Khalid, whose work explores themes of diaspora, identity, family, and home. Lali is currently a visiting artist at the School of the Art Institute of Chicago. Originally from Pakistan, she lived in Marquette for several years before moving away for personal and professional reasons. Lali has returned to Marquette every 10 days for three nights to visit her son for the last 6 years. These regular journeys—both emotional and physical—form the basis of her photographic contribution to this project. Her images are shaped by the complexities of motherhood, migration, and a legal system that has redefined her relationship to place and family. The Upper Peninsula becomes not just a site of natural beauty, but a terrain marked by longing, distance, and quiet resilience. Lali holds a BFA in printmaking from the National College of Arts in Lahore and an MFA in Photography from Pratt Institute. She moved to the United States in 2011.

Who Is this Book for and Why Write this Book?

It is my hope that this book will resonate with a wide range of readers.

This work serves as an oral history of those who arrived from abroad into Marquette county relatively recently and have established themselves in important and visible roles. The group of participants who made this project possible will gain a permanent record of their migratory journey.

I hope that my academic colleagues will see this work as filling a gap in literature about the livelihood of international professionals who live and work in remote, small cities in the United States as it is my belief that stories of professional immigrants often go unnoticed.

I hope that my students will find "Moving UP North" applicable to their academic journey. The book indirectly addresses the topic of capitalism and immigration flows into small towns and rural areas that some may find intriguing to think about.

Furthermore, there is a gap in literature about the individual experiences of international professionals who live and work in small towns where they may be the only representatives of their countries, where access to fresh ethnic food is limited, where religious services may be unavailable, and where being foreign is "visible" in the community.

Finally, I believe that my fellow international geographers will find this book applicable to their own communities. Limited diversity in small towns is certainly not unique to the United States. In fact, it is relevant to rural areas in other countries that attract immigrants and refugees. I believe readers in Europe, especially in the UK, Italy, or Spain, will find this book relevant to the reality of their own hometowns. Readers in new immigrant destinations in Central and Eastern Europe should find this topic very thought-provoking as well. Newly emerging immigration hubs such as Poland attract diverse immigrants (from Georgia, Moldova, Vietnam) and Ukrainian refugees who settle in small towns and rural areas due to the availability of jobs. These immigrants face similar challenges such as lack of diversity, religious services, and standing out in their host communities.

I would like to underscore that I am writing this book aware of the fact that many topics in the current American discourse are highly politicized or may be perceived as making a political statement, both politically liberal and politically conservative. That is not the intention of this book. The interviewees selected for this work were selected by the author without any knowledge about their experiences of living in the United States or in Marquette, MI, specifically. I am hoping therefore that these stories are analyzed without a political lens and that all readers will find these stories as valuable sources of first-hand knowledge about how complex migrant experiences can be. I also hope that readers will find that the history of where and in what structure individuals grew up has directed the migrants' experiences in the United States almost as equally as any qualities of American life itself. What I am hoping the readers will also reflect on is the fact that beyond the professional triumphs and the overt establishment of new lives, many of these migrants endure profound personal struggles—feelings of isolation, homesickness, cultural dislocation, and moments of vulnerability as they navigate unfamiliar social landscapes. In the stories featured in this work, the unfamiliarity of landscapes is often broken by the most common concept: a single person or a small group of individuals that connects around shared experiences, human nature, or simple empathy. Those intimate human connections are often significant factors that place individual immigrant experiences on new trajectories. And while this book features many optimistic and happy stories, as readers review these stories I hope they will keep in mind that individual interactions and actions can just as profoundly make those immigrant trajectories quite negative.

References

Bachelard, G. (1994). *The poetics of space* (p. 29). Beacon Press.

Baxter, A. M., & Nowrasteh, A. (2021, August 3). *A brief history of U.S. Immigration Policy from the Colonial Period to the Present Day*. Policy Analysis no. 919, Cato Institute. https://doi.org/10.36009/PA.919.

C.B. Rao Oral History Transcript. (1997). Upper peninsula digital network. Central Upper Peninsula and Northern Michigan University Archives. Red Dust Oral History Collection. Immigration. https://uplink.nmu.edu/islandora/object/nmu%3A135567?search=Rao. Accessed on 5/14/2024.

Camarota, S., & Zeigler, K. (2024). *The foreign-born share and number at record highs in February 2024. A look at the size and growth of the immigrant population in the United States*. Center for Immigration Studies. Articles last accessed on November 30, 2024. https://cis.org/Report/ForeignBorn-Share-and-Number-Record-Highs-February-2024

Cresswell, T. (2011). Mobilities I. Catching up. *Progress in Human Geography, 35*(4), 550–558.

Hass, H., Castles, S., & Miller, S. J. (2020). The age of migration. In *International population movements in the modern world* (6th ed.). Guilford Press.

Harzing, C., Hoerder, D., & with Gabaccia D. (2009). *What is migration History?* Polity Press.
Hamnett, C. (1994). Social polarisation in global cities: Theory and evidence. *Urban Studies, 31*(3), 401–424. https://doi.org/10.1080/00420989420080401. Original work published 1994.
Kusek, W. 2016. Polish(ing) my identity. Book chapter in Navigating geography and self: Transnational geographers disrupting boundaries in a global age. Alan P. Marcus (Eds). Lexington Publishing.
Kusek, W., & Wise, N. (2020). *Human geography and professional mobility. International experiences, critical reflections, practical insights. Routledge studies in human geography*. Routledge Taylor & Francis Group.
Magnaghi, R. M. (2017). Upper Peninsula of Michigan: A history paperback—June 4, 2017.
Malinowski, J. and D. Kaplan. 2012. Human geography. McGraw-Hill Higher Education.
Marcus, A. P. (Ed.). (2016). *Transnational geographers in the United States: Navigating autobiogeographies in a global age*. Lexington Book.
Michigan League for Public Policy. (2024). *A snapshot of Marquette County immigrants*. https://mlpp.org/2022immigrantsnapshot/Marquette.pdf.
Milton, C. (2024). *NMU Spring 2024 Commencement Speech*. PBS. https://www.pbs.org/video/nmu-spring-2024-commencement-uzm7i9/#:~:text=Special%20%7C%202h%2044m%207s%20%7C,commencement%20speaker%20is%20Milton%20Chen.
Obituary of Dr. Miodrag Bata Georgevich. (2004). Found at https://obits.funeralinnovations.com/obituaries/view/374449/2/. Retrieved on August 9, 2024.
Ray, B. (2003). The role of cities in immigrant integration. *Migration Information Source*. Feature. October 1, 2003. https://www.migrationpolicy.org/article/role-cities-immigrant-integration.
Sassen, S. (1985). Capital mobility and labor migration: Their expression in core cities. In M. Tinkerlake (Ed.), *Urbanization in the world economy*. Academic Press.
Sassen, S. (1991). The Global City: New York, London, Tokyo. Princeton University Press, Princeton.
Travel Marquette. (2024). https://www.uptravel.com/things-to-do/. Accessed on December 13, 2014.
Tuan, Y.-F. (1974). *Topophilia: A study of environmental perception, attitudes, and values*. Prentice Hall.
Tuan, Y.-F. (1977). *Space and place. The perspective of experience*. The University of Minnesota Press.
Wright, J. (1947). The place of imagination in geography. *Annals of the Association of American Geographers, 37*(1), 1–15.

Chapter 2
Zosia

Introduction

Zosia Eppensteiner is the CEO of the Community Foundation of Marquette, MI. Zosia was born in Gdansk, Poland. She holds two masters degrees—one in Math and one in Sociology—from the University of Gdansk. Zosia moved to Marquette in June 2005 and worked at the Wells Fargo Bank and First Merit Bank as a Customer Service Manager. She was also employed part-time as contingent faculty at Northern Michigan University where she taught evening courses in math. In 2013 she joined the Community Foundation as Director of Finance and was promoted to CEO in 2020 when the former CEO retired.

Early Childhood and Influence of Place

Zosia was born in Gdansk in 1981 just when Poland introduced Marshall Law in reaction to growing opposition to the dictatorial, communist, and Soviet-controlled government in Poland. This was a time in Polish history when average citizens in Poland were not allowed to travel abroad, where there were restrictions on owning or even accessing Western goods, and the population uniformly shared a desire for freedom.

Zosia grew up in an efficient, communist-era apartment in Gdynia providing 600 square feet of living space. Her home was small, and like most families who were existing in a communist economic system, her family had modest financial means. That said, the family maximized lifestyle and cultural opportunities available to the general public in larger Polish cities. She recalls playing with her sister and kids in the neighborhood, walking to a nearby convenience store, and spending a lot of time outdoors. She recalls walking freely in the community and taking public

W. A. Kusek, *Moving UP North*, https://doi.org/10.1007/978-3-031-95701-7_2

transportation to the movies. As a well-interconnected urban area, Zosia's home city afforded a lot of freedom for children and teenagers. The area was safe and neighbors knew each other well, and both the walkability and ample public transport meant that Zosia was not confined to her family home. Zosia's parents did not hold college degrees—they both had high school diplomas which at the time was a common level of education for many Poles. As the political and economic system in Poland transitioned into a democracy and a capitalist, free-market economy, Zosia's parents remained in manufacturing. Her father worked in the Gdansk Shipyard and her mother, the main breadwinner, worked as an accountant in Polipol—a factory that mostly employed people with disabilities. Zosia recalls visiting her mother's workplace and interacting with the employees who had a variety of disabilities, an experience that now she sees as formative to what her professional role is at the Community Foundation in Marquette.

Although Zosia's parents did not graduate from college, they were very aware of what was happening around the world. Due to Poland's historical vulnerability to global conflicts and developments, many people in Poland felt the need to closely follow global events to be prepared for consequences or outcomes that may impact their life in Poland. Zosia's father loved history and also shared many stories from his childhood as well as historical records with his young daughters. The parents considered educational attainment as a key for success and made sure that their daughters were good students and had every educational opportunity the family could afford. The family considered themselves people of high culture and celebrated the family's contributions to Polish cultural heritage. Zosia's grandfather spent 5 years in a Soviet prison and wrote a book about his traumatic experiences. Her grandmother was a dentist who traveled to Nigeria to provide dental care for the local community. People in her family were courageous and strong, worked hard, and knew a lot about the world at a time where the communist government was restricting access to information. While Zosia's parents could not travel much abroad due to the realities of existing in a communist country behind the Iron Curtain, the family made it a priority for their daughters to have a better life and to experience the world. When the Iron Curtain ceased to exist, Poland became a democracy and Poles were finally free to travel internationally.

New opportunities became a reality for Zosia when Zosia's mother reconnected with a friend who was a math professor at a university in Charleston, SC. Zosia's parents were invited by their friend to visit her in the United States but instead decided to send their two daughters to allow them to experience summer in the United States. Zosia was 13 and her sister was 14 when the two of them boarded a plane, navigated a connecting flight in Philadelphia, and arrived in Charleston, SC, on their own. They spent two months in the United States visiting the South and traveling to Disney World. They also spent time on the university campus where their mother's friend worked. Zosia attended summer lectures in precalculus and she spent a lot of time in the university library where she was translating songs from English to Polish. For 13-year-old Zosia the American experience was truly transformative and changed her personal trajectory. Before the trans-Atlantic trip Zosia self-identified as a "homebody" and was less comfortable with the idea of exploring the world. Visiting the United States gave her confidence in her abilities and a strong belief that she could accomplish anything she set her mind to. She saw many opportunities and understood that she can make it anywhere she moves. After two transformative summer months in the United States, Zosia and her sister returned to Poland. Zosia went to high school and completed an International Baccalaureate program. During the break between her high school graduation and the start of her college education, Zosia decided to reconnect with the United States via a summer work program offered by Camp America. She arrived in Crystal Falls, MI, where she worked as a camp counselor. This is where she met her future American husband Eric who was also a camp counselor. After the summer in Crystal Falls, Zosia returned to Poland to begin her college career. Zosia kept in touch with Eric for five years as she was completing her university education in Gdansk. Zosia and Eric travelled between Poland and the United States to see each other. Zosia was visiting Eric in the summer in the United States and Eric was coming to Poland for shorter visits during the academic year.

Migration to Marquette, MI

Zosia moved to Marquette, MI, in June of 2005 and married Eric in September of that year. She arrived in a familiar place as she had visited the Upper Peninsula several times prior to moving here permanently. She moved to the United States for love so she felt she was moving to Marquette to be with someone who was already close to her heart. As such, Zosia did not experience feelings of isolation. She also had an existing network of friends and Eric's family was very welcoming and supportive. Although the Upper Peninsula of Michigan is very different from the South that she first visited as a 13-year-old child, she quickly fell in love with this location and the people.

Soon after arriving in Marquette, Zosia began working at a local bank, she supplemented her bank work by helping out in a flower shop, and she taught a math class at Northern Michigan University. Zosia forged new friendships, and Eric introduced her to local bars and familiarized her with music events in town. While Zosia was fascinated with her new home, Zosia vividly recalls experiencing some level of culture shock. The architecture and urban layout in Marquette was vastly different from Gdansk, Poland. Zosia was used to historic downtowns, beautiful historic churches, and a vibrant and social Old Town. In stark contrast to the historic grandeur of Gdansk, Marquette's modest architectural landscape underscored for Zosia not only a physical but also an emotional distance from the cultural richness of her past. Additionally, while Zosia was used to walking and interacting with people on the street, in Marquette it seemed that everyone drove their cars to their destination. Food looked similar to what it looked like in Poland, but had less taste and was more processed, and people seemed to treat food as a daily necessity rather than a language of love or artistic expression. Interestingly, the unifying factor was Lake Superior that reminded Zosia of the Polish coast on the Baltic Sea *(please see* Image

2.1). She loved going to the beach and appreciated that, as opposed to the crowder and touristy Baltic Sea towns, Marquette's beaches seemed more natural and were much less crowded. While living in Marquette, Zosia learned how to snowshoe and became a fan of cross-country skiing. Moving from the city to the rural but also outdoorsy Upper Peninsula, Zosia gained a bigger appreciation for the outdoors and she started integrating outdoor activities into her well-being routine. Zosia's strongest connection to Marquette formed when she became a mother to her son Max.

Home Away from Home: Blending Local Support and Personal Qualities to Navigate Migration Challenges

I found Zosia's history fascinating, as her migrant identity, perceptions of Poland and Marquette evolved over time as her personal situation changed. Today, Zosia admits that the level with which she embraced her immigrant identity increased after she got divorced from her husband Eric. The divorce intensified Zosia's focus on understanding her own, personal identity. For many years after arriving in Marquette Zosia did not identify as an immigrant at all. She was a homemaker, a mother, a wife, and a professional. After life-changing events she understood that her immigrant identity is strong and enriches her work in the Community Foundation. She considers herself privileged to have had the experiences and opportunities her life presented and strongly feels that her privilege and experiences come with a

Image 2.1 Lake Superior. (Photography by Lali Khalid)

significant degree of responsibility. She feels responsible to help others who are less fortunate than her. Those people include individuals who are homeless, face addiction challenges, or have disabilities. Zosia feels these deserve support and a safety net that only a community organization can provide.

Zosia admits that as an immigrant in Marquette she received a lot of support from her ex-husband, her ex-mother in law, and their family and friends. When she arrived she felt welcomed, supported, and had wonderful people around her who made sure she could create her home away from home. She recognizes that not all immigrants have that experience and feels that her personal arrival story added to her list of personal privileges. She also emphasizes that it is often the perceived support rather than actual tangible support in the form of housing or finances that immigrants truly need to feel part of the community. As Zosia states, immigrants want to build local relationships and they do not want to feel as outsiders. Signs or actions that show acceptance by locals are particularly appreciated.

As Zosia recounted her migrant journey, she tried to identify personal qualities and experiences that she feels defined her initial migrant experience as well as subsequent career opportunities. First, Zosia understands that educational and social capital have given her access to opportunities that not all immigrants and not all local residents can aspire to achieve. Zosia arrived at Marquette already highly educated. She was fluent in English. She was experienced with international travel. She had a strong work ethic. She also felt she made decisions thanks to countless influential individuals and mentors from her past. Most importantly, her family and upbringing in a country that almost overnight completely swapped political and economic systems equipped her with the confidence and ability to adapt to a new environment. Zosia also feels she grew up in a reality and in a family which instilled in her a belief that a woman can be successful anywhere. She also came from a family that was supportive of her decisions. Her parents always believed in her, and when she moved to Marquette, as a result of her upbringing, she also believed she could be successful here as well.

After carefully inventorying all the factors that Zosia feels contributed to her experiences and success, Zosia carefully transitioned to parts of her immigrant experience that were less pleasant and more challenging. Zosia emphasizes that immigration in general is a drastic change in the life of a person who decides to move countries. It is a catalyst to exponential growth but it is also isolating. An immigrant often feels misunderstood by the people they are leaving behind and the people in an immigrant's new locality. Neither group can practically relate to the immigrant's experience or perspectives. Zosia stressed that as a human being and especially as an immigrant, a person changes, learns, and redefines their reality; an immigrant is constantly exposed to new experiences that change their personality, perceptions, and understanding of the world.

That drastic change may affect personal relationships. Zosia fully recognizes that her marriage did not survive for many complex reasons but one of them was the change in her own identity. She was no longer the person that first arrived in Marquette in 2005. Her priorities, goals, and aspirations changed. Zosia also highlighted that immigrants often have a lot of guilt related to leaving behind their aging parents. Many people who never experienced immigration do not understand the guilt, the questioning of one's choices, and the dual identity that are a common experience for migrants.

When asked to elaborate on details related to having a dual identity, Zosia emphasized that many migrants transition among two identities—one is a result of experiences from two places combining into one person. Zosia is one person defined by both Poland and the United States, by both Gdansk and Marquette. That said, Zosia also understands and can blend into one of those realities at any point, and in many cases only half of her experiences and identity is relevant to the people around her. This leads to constant changes and fluctuations on what an immigrant experiences and how they define themselves. By being an immigrant, a person expands their network; knows more sights, sounds, smells, and tastes; has broader perceptions of the world; and can place many norms and rules in the local context of how they originated or why they are locally important. Having these experiences, this knowledge, and these perspectives can lead to new opportunities and help influence people, processes, and communities in amazing ways. On the other hand, as Zosia indicates, that identity is not permanent, and depending on the environment or location, an immigrant constantly questions who they really are and can switch identities from one to the other. When asked about examples, Zosia indicated that this manifests in both subconscious and obvious, overt ways. One example Zosia cited is related to the most easy to perceive indicator—fashion and looks. Zosia highlighted how when she visits Poland, her consistent fashion choices are seen as plain, bland, or modest, and those same clothes in Marquette earn her compliments for being stylish. So when visiting Poland, over time, she adapts her fashion sense and fashion choices. Zosia also feels she adapts her rules of social engagement depending where she is. In the United States, Zosia would never ask a neighbor for a cup of sugar if she ran out in the middle of preparing a cake for a celebration, while in Poland she would knock on her neighbor's door without hesitation. Interestingly, Zosia pointed out that even the simple act of walking is different depending on the context. In Marquette, Zosia walks for leisure and relaxation. In Poland she walks to get somewhere, to reach a specific destination.

Anchoring Around a Purpose

When taking into account all the abovementioned deep challenges to her personal identity, Zosia understands that as an immigrant she is ultimately from two places but also in a way she feels she is "neither from here nor there." That feeling of dissociation and lack of belonging becomes evident when Zosia considers broad geographic locations—Poland, the Untied States, Gdansk, or Marquette. But it disappears when she discusses the Community Foundation. As Zosia recounted, it took her many years to establish a professional career in Marquette. She worked different jobs and it was not until Gail (a friend of her ex-husband's family) presented her with the opportunity to join the Community Foundation. Gail believed in

Zosia and encouraged her to dream big. Zosia considers Gail a "third mother." The two women built a community organization that today helps hundreds of people. The organization offers a variety of services, grants, and scholarships. They fundraise for a variety of community projects such as building the recently opened inclusive playground in Marquette. For Zosia such projects help her close a metaphorical circle of events and traditions. Zosia sees that her mission started with her own mother who worked with disabled employees and supported underprivileged communities in Gdansk. Today, Zosia also has an opportunity to lead a community charity that helps Marquette community members in need and through this action connects geographies and her own split identities. Despite these conflicting personal identities, it is thanks to the Community Foundation, the work she gets to do, and the people in her everyday life that she considers Marquette home. She simply feels she is an important member of the community. It is through marking a difference locally that Zosia feels her entire upbringing and story all comes together for a very particular reason. Zosia's reflections on her dual identity—rooted in both Poland and the United States—can serve as a preview for later chapters in this work where other immigrants similarly narrate their struggles and triumphs in reconciling past and present.

Chapter 3
Hema

Introduction

Hema Tripathi is a licensed physical therapist and board-certified orthopedic clinical specialist (OCS) at Advanced Orthopedics and Plastic Surgery in Marquette, Michigan. She holds a graduate degree, Master of Science in Physical Therapy from the College of Applied Education and Health Sciences, from a leading academic and research institution in India—the Manipal Academy of Higher Education (MAHE). She is married with 2 children. The oral history of her husband Ritesh's migratory journey is also featured in this book. Please note that Hema's husband, Ritesh, offers his perspective on their shared migration experience in the next chapter.

Early Childhood and Influence of Place

Hema was born in Palakkas in the state of Kerala in India. Her parents were domestic migrants who moved from the western part of India to the southern part of the country with a goal of establishing a business. The family lived in a house that was located on a big piece of property nearby a small town. The family owned a sawmill and her father processed wood logs. Hema was part of a bigger community of people who worked together and supported each other. Her grandfather employed a lot of people and the family had a network of friends. Culturally, there was loyalty and a feeling of responsibility towards one's community; hence, when possible, the family shopped in stores owned by people within Hema's community network. Hema was always surrounded by familiar faces. Culturally, in India, the family and community network take responsibility for protecting girls from threats of the outside world. Girls are surrounded by family and familiar people and often restricted from

W. A. Kusek, *Moving UP North*, https://doi.org/10.1007/978-3-031-95701-7_3

venturing out into the outside world without familiar family members. Hema, like other girls in her family, was not given much independence. That restrictive protectiveness differed greatly from Hema's brother who grew up with much more individual freedom.

Hema grew up in a multigenerational home where the idea of personal privacy was a foreign concept. She shared a room with her brother, the family owned 1 TV, and they mainly ate meals at home within their family circle instead of frequenting restaurants. Visiting restaurants was so rare for Hema; she vividly recalls ordering her first pizza outside of her home at the age of 16.

As Hema was describing her childhood and upbringing, Hema found it important to identify elements of her upbringing which may be unique to life in India. Hema highlighted, for example, that in India baking is not a consistent part of the culture, so Hema's home did not have an oven. Bread was something you would always buy from a local bakery. Hema also recalls not attending birthday parties or not having scheduled school playdates. Kids just played outdoors with other children who lived nearby.

India is a diverse country with different cultures and traditions. Moving from one region of India to the next is often similar to moving from country to country. When Hema's family moved to Kerala, they needed to adjust and assimilate to a new way of life within their own country, if they wanted to become members of their new community.

Linguistically, India has 22 languages that are officially recognized by the government. It is not surprising that Hema grew up in a family who spoke a different language than the one spoken in Palakkas, and many of her family's traditions and cultural rituals were different from those of Hema's neighbors and friends from school. That experience with transitions helped Hema with her eventual move to the United States as she was already used to making adjustments to a new place, a new culture, and a new way of life. When Hema's family moved within India, she had to learn folk dances that were part of southern India's culture. Her family also had to learn how to celebrate Hindu festivals in the south.

Living, migrating, and learning in India—especially as a former British colony—meant that Hema's experiences with various cultures, languages, and even educational styles was acquired early on in her life. Hema's parents wanted her to receive the best possible education. Due to that goal, Hema attended a Catholic school—which was part of a system of schooling established during British rule. Hema's parents also supplemented her schooling by hiring a tutor. Hema pointed out that the tutor was also Christian and had a tremendous influence on Hema's educational journey—even today she often thinks of him and holds him in very high regard. Attending Catholic school and interacting with Christian tutors did not limit Hema's religious expression. Even though Hema attended a Catholic school during the day, in the evening she was visiting a Hindu temple. For Hema, transitioning between different religions and traditions was seamless. Although Hema did not consciously think of cultural and religious diversity at the time, she realizes that due to her upbringing, she has become familiar with different religions and was able to learn and speak 7 languages. English was the language of instruction at her school which became particularly useful when she moved to the United States later in life. Interestingly, although the language of instruction in school was English, it was not the language spoken among school friends. Hema spoke her mother tongue at home with her family, a different language with her neighborhood friends, and English with teachers and at school. Later she learned Hindi and additional 2 more languages. Hema attended a coed school for the first few years of her education and then was sent to an all-girls school. She was expected to perform academically, and although she grew up completing some chores at home, her main job was to study and succeed academically.

Hema's close relationship with her family and upbringing in a controlled environment lasted for most of her youth. It was not until college when Hema moved away to a different town, but even then, her move was planned and she was not permitted to move away too far from her family. Hema dreamt about attending dental school but there were no dental schools in her vicinity, so she instead chose to enroll in a Physical Therapy major at a university which was located in a town where her extended family lived. Hema was sent to live with her uncle and his family and was studying to be a physical therapist, a profession she knew very little about and one that was not in her immediate scope of interest at the time.

While in college, Hema met Ritesh, her future husband. He proposed to her when they were finishing up their studies. While Hema and Ritesh eventually got married, at first they had to work through family dynamics, as the marriage proposal was not accepted by Hema's parents.

Marriage as a Social Construct, Class System, and Caste

Marriage in India is a social construct that is influenced by a variety of factors, including religion, caste, class, region, and a family's attitude towards the bride or groom.

The class and caste system in India differs from the understanding of class in the western world and there are many rules and traditions associated with how people in India perceive themselves and others based on their socio-economic status.

Although recent studies suggest that Indian culture is trending away from traditional arranged marriages, families still play an important role in influencing who their daughters and sons will marry. Only a minority of marriages happen without the blessing of the parents. In Hema's case, her parents did not approve of Ritesh' marriage proposal to Hema as they did not believe he would be able to provide a good enough quality of life for their daughter. Hema's parents viewed their family as existing and living in a superior cast to that of Ritesh, and they wanted a husband for Hema that they felt could guarantee a high socio-economic future for their daughter. After completing her university degree, Hema returned to live with her parents and her family started presenting different young men to her as potential marital candidates. Hema recalls a lot of drama associated with the search process for her future husband. She remained firm in her feelings, however, and finally told her family that she chooses to stay single forever if she cannot marry Ritesh. She was not going to marry him without her family's blessing because she would not want to lose contact with her family—an inevitable outcome of a decision to marry against her parents' will.

It took years for Hema's mother to finally approve of Ritesh. Five years after the initial marriage proposal, Hema's mother sent Hema's uncle and brother to visit the home of Ritesh's parents to discuss the marriage and wedding. The couple finally married and began their family life in India. Hema and Ritesh worked at a college where they were teaching classes and Hema was completing her graduate school. They had a stable and happy life and they were not considering leaving India until Ritesh visited his brother who lived in California. He passed a variety of exams to qualify for American jobs, secured a job at the Marquette General Hospital, and moved to Marquette in October 2009 while Hema stayed in India and lived with her parents in law for a few months.

Migration to the United States

While Ritesh was setting up life in the Upper Peninsula of Michigan, Hema was preparing to join him in the near future. She arrived in Marquette in February 2010 in the midst of a UP winter. Ritesh prepared their rental apartment for her arrival to the best of his ability and means, but it took the couple several months to save enough money to furnish their new home. As the couples' careers progressed—they secured all the required permits to permanently work in the United States—they also advanced personally and started a family. Their two children were born in Marquette. As an immigrant to a place so dissimilar from her home, Hema stressed that her experience with pregnancy in Marquette helped her discover and appreciate the "spirit of the people in Marquette." Her coworkers offered a lot of emotional support which was extremely important for an immigrant woman who did not have her mother nearby. Hema's friends organized a baby shower for her and gave her many baby accessories that Hema needed. They also advised her about car seats and strollers, and a co-worker even gifted her a breast pump. Hema recalls that episode as a turning point—not only did it help her forge a supportive network in Marquette, but it also subtly revealed the bittersweet duality of migration: the simultaneous joy of newfound community and the quiet ache of leaving behind a life steeped in familiar traditions. That said, Hema felt that her life mattered to others and that she again had people with whom she could share her experiences. The following chapter that discusses the experiences of Ritesh provides a parallel perspective of how Ritesh also navigated feelings of isolation, identity, and community formation.

Hema also recalls the process of purchasing their first home as an event in her life that confirmed her intuition about the fact that Marquette is the place where she wants to raise her family. The couple was ready to purchase a home in Marquette but Hema had very specific requirements in mind that were not easy to find in a small town. She wanted her house to face the East which is important in Indian culture. She wanted the backyard to be fenced-in so her kids can safely play while she is taking care of household chores. For that reason, it was also important to have a kitchen window that would face the backyard. Hema also had a particular architectural style in mind: a ranch-style home which would allow her children to have their bedroom on the same floor as Hema and her husband, but also a home without stairs for Hema's parents who would be visiting from India. Hema recalls that her local UP realtor not only did not question or challenge these requirements, but despite limited choices searched for the right home until it was found. Finding the home was not enough however. And this is where another quality of Marquette came through—despite being a remote location in Michigan's Upper Peninsula—a place with relatively little diversity and exposure to the outside world—the previous owner of Hema's home specifically chose them as the next owners. Despite receiving multiple offers on the home from other buyers, the previous homeowner chose to sell the home to Hema and Ritesh because they were raising a new family. The needs of a new family were personally important to the homeowner, who met Hema and Ritesh and learned that Hema was pregnant and due soon. In fact, the property seller even thoroughly cleaned the house for the new family and prepared it so well because she wanted to make sure that the pregnant Hema would not need to spend the time cleaning it. Hema said that when she moved in, every corner of the house was spotless. Hema again felt the spirit of the Marquette community, how nice and supportive the people are, and she again felt that this is the community she wants to call home. As Hema mentions, that was not even the end of the story, as a few months later the previous owner of the house returned to offer advice to Hema and Ritesh on how to best care for the plants in their backyard—a knowledge they did not possess and found incredibly helpful.

Home Away from Home: Importance of Culture, Tradition, and Religion

When immigrants arrive in the United States (a statement that likely applies to immigrants in other countries as well), they often rely on and appreciate having a support network to help them work through the initial transition to a different way of life. In large urban areas many immigrant groups can rely on their ethnic community to provide some level of support, but in rural areas like the Upper Peninsula of Michigan migrants must rely on the local population and the local infrastructure to offer support to newcomers.

Hema once again recalled her experience as a new mother in the United States and in Marquette specifically as an episode where that support was particularly needed. She was not just a new mother; she was also working full time. At first, she felt guilty about sending her children to daycare. But for a professional immigrant on a H1B visa (a permit requiring immigrants to maintain full-time work status), it was impossible for her to stay at home with her children. Hema emphasized that despite initial reservations, she had a wonderful experience with the caregivers at the daycare who were helpful and supportive and also understanding of the challenges of navigating responsibilities of motherhood while maintaining full-time employment.

Support with her pregnancy and a reliable childcare infrastructure were important contributors to Hema's feeling of belonging in Marquette. Hema indicated, however, that she desired to understand and experience local life in the UP and it was a friendly "Yooper" that really enabled her family to feel at home in Marquette. Hema's friend Jesse and his mother Nancy were the family's ultimate catalysts in experiencing and learning the way of life in the UP. For example, Jesse and Nancy invited the Indian family to their first Thanksgiving dinner and since then Hema and Ritesh always plan to celebrate this American holiday with Jesse and his family. The families became true friends and Hema emphasizes that a lot of Yooper traditions that she and her family learned are thanks to Jesse' guidance. From Jesse, Hema and Ritesh learned how to enjoy the natural beauty of the Upper Peninsula and experienced camping, hiking in the woods, or preparing outdoor bonfires. Jesse also taught Hema's family how to mountain bike and he often takes Hema's children to ride on Marquette's fairly extensive mountain biking trails. Hema emphasizes that it is particularly important to identify and befriend a local person who can serve as a liaison to the local culture and community activities. As Hema and Ritesh aspired to truly become practitioners of local culture and activities, they explored other local opportunities thanks to Jesse. Canoeing together on the AuTrain River became an annual tradition for Hema's and Jesse' families. They also go berry picking in the summer, they drink cider in the fall, and they cut a Christmas tree together with Nancy in the winter. Although Hema and Ritesh did not grow up with a Christian belief system, they adopted the tradition of exchanging small gifts for Christmas and they learned about the importance of Christmas ornaments. What is interesting and particularly encouraging is that the cultural exchange between Hema and Jesse's families is not unidirectional. While living in Marquette, the Tripathis also celebrate Indian festivals. Hema shared the tradition of Rkhi Festival—a celebration of the relationship between a brother and sister. The Tripathis have 2 daughters and they rely on Jessy—their Yooper friend—to fill in the role of a brother for the purpose of celebrating this festival. They modified the tradition because Hema did not want her daughters to miss out on this important and fun festival.

Incorporating Jesse's family in the practicing of an important Indian holiday is a perfect example of how immigrants to rural communities, once able to forge trusting, local relationships, can truly enrich the lives of people locally while at the same time learning and embracing local traditions.

Chapter 4
Ritesh

Introduction

Dr. Ritesh Tripathi is a board-certified clinical specialist in Neurologic Physical Therapy and an owner of Tenet Therapies in Marquette, MI. Dr. Tripathi is a graduate of the Manipal Academy of Higher Education in India and also holds a doctorate in Physical Therapy in Neurology. He grew up in New Delhi, India. Dr. Tripathi arrived in the United States in May 2009 and he permanently moved to Marquette, MI, in October 2009. He is married to Hema and the couple have two children (for Hema's migratory journey and experiences please refer to Chap. 3). Presenting the stories of Ritesh and Hema is a particularly insightful and interesting endeavor, as their stories are heavily influenced by cultural norms in India. I also feel that Ritesh's story amplifies more the professional dimensions of migration, while Hema's earlier account offers a complementary view on the emotional and cultural transformations that mark their shared experience. As professional immigrants to the United States they are people who together provide both a perspective of the migratory journey to the United States and also interesting insight into the reality of growing up within the Indian society, how men and women are raised, and how emigrating from India into another country can effectively challenge Indian societal norms and place people's futures on a new, fresh trajectory.

Early Childhood and Influence of Place

Dr. Tripathi was born in New Delhi in India. New Delhi is a political, cultural, and economic center of India, a mega city that houses almost 34 million people. The city is a melting pot of domestic and global cultures. The city is composed of many smaller communities and attracts a diverse group of people from all over India as

W. A. Kusek, *Moving UP North*, https://doi.org/10.1007/978-3-031-95701-7_4

well as from all over the world. Dr. Ritesh Tripathi emphasized how he grew up surrounded by a vast diversity of people and cultures. His experience of diversity was not only related to the ethnic, religious, and linguistic diversity within India itself, but also attributed to the diversity of immigrants who settled in India and were attracted to India's main global city. He recalls growing up and seeing flags that represented countries from around the world, tasting ethnic foods, and visiting western restaurants all while being heavily influenced by the domestic diversity of India. Ritesh grew up in a small community that was part of a vast, big city. He grew up in a small house without a yard on a cul-de-sac. His family spoke Hindi. He also spoke Hindi at school; he was taught Hindi culture and Hindi traditions. That said, his house was located near a Sikh temple (another religion in India). He recalls very positive memories related to interacting and playing with Sikh children and he recalls how these people from different cultural traditions respected each other. He remembers participating in a variety of Sikhs events where he helped to clean up after the events and dined with his Sikh neighbors in the gurdwara (a place of worship and communal dining hall). All the children played together and deeply cared for one another. Despite the variances in cultures and traditions, all these people seemed to respect one another and cared for one another. As such, religious diversity was a normal part of Ritesh's upbringing. He was exposed to Christian and Muslim religious traditions as well. He was taught to respect Christian and Muslim holidays and he remembers that children loved celebrating blended holidays. This experience growing up made it easy and natural to understand and respect diverse populations later in life. While in college, Ritesh rented living accommodations from a Catholic family, and when he moved to the United States, he found the American diversity of people, traditions, and ways of life as normal and expected.

Dr. Tripathi was exposed to diversity throughout his life, and similarly to other migrants interviewed in his book, he recalls that his exposure to diversity often manifested itself in traditions related to food. Food culture was often at the center of community celebrations and food provided a connecting point for everyone. Neighbors shared and exchanged food and celebrated religious events with food. He joked that he grew up a "foodie" before this term was even coined.

Ritesh explained that for cultural reasons his family did not consume meat, eggs, fish, and alcohol. A cow in India is considered a blessing and celebrated like a second mother to people, so beef specifically is not consumed by many people there. Ritesh interestingly highlighted that the American restaurant chain "McDonalds" offers a different menu in India that features mostly vegetarian items and some chicken dishes as well. McDonalds in India also serves potato patties and cheese patties. Ritesh's story highlighted that immigrant experiences are not purely about learning American cultural anchors and symbols, but sometimes immigrant experiences involve relearning what certain American exports mean in a local American setting. A McDonald's restaurant in India is a very different restaurant than it is in its original birthplace in the United States.

The main focus of my interview with Ritesh was on his relationship with Hema (featured in Chap. 3) and how Ritesh's journey to the United States helped him distance himself from rules and norms existing in Indian society. India functions based on the caste system—a social classification that dictates social, economic, cultural, and familial roles. People are born into a particular caste or "class" and caste mobility is very restricted if not impossible. The caste system creates rules and perceptions in India that are foreign and difficult to understand in the United States and in the Western world. One important outcome of the caste system is that it is difficult to marry people outside of one's caste, and a person's caste also influences which colleges or professional opportunities they can access.

Ritesh came from a well-educated, loving family. His mother held a bachelor's degree in math and his father held a bachelor's degree in science. Because education was very important for his parents, they made sure that Ritesh received a solid educational foundation. Ritesh attended a charter school in India and his family aspired for Ritesh to follow the trendiest career paths in India and become either an engineer or medical doctor. The challenge was that competition for these majors at Indian universities is extremely high. There are several thousands of students competing for each open spot, and many seats are reserved for students coming from specific casts. Although Ritesh wanted to enroll in a medical school, the combination of extreme competition and the preselection of some open spots for students from lower castes precluded Ritesh from attending university in one of his desired majors. Instead of medical school, Ritesh enrolled in a Physical Therapy program. His academic path created a path for his personal life. In college Ritesh met the love of his life—Hema. By the end of their college studies Ritesh proposed to Hema.

As also mentioned by Hema in Chap. 3 the caste system created barriers for their marriage. Hema's mother agreed for Ritesh to marry Hema after 5 years after he had proposed and many family negotiations. The couple lived a comfortable life in India. They both worked at a college and did not consider emigration until Ritesh visited his brother who lived in California and learned about economic opportunities in the United States.

Migration to the United States

Ritesh arrived in California on a tourist visa in May 2009. He came to visit his brother who is an IT professional. During his trip to California Ritesh realized the breadth of economic opportunities in the United States. He decided to complete a variety of paperwork required to secure legal employment in the United States. He studied for and passed several exams that allowed his professional degree to be recognized in the United States which in turn allowed him to seek employment. During this whole process Ritesh returned to India to apply for a work permit in the United States at an American Embassy in New Delhi. Dr. Tripathi applied for 600 jobs in the midst of the economic recession and received just 3 interviews. One of the interviews was with the Marquette General Hospital in Marquette, MI. Dr. Tripathi was very excited to interview for a job in Marquette because the State of Michigan recognized his Indian education and accepted college credits from his Indian university.

Dr. Tripathi arrived in Marquette on October 26th, 2009, while his wife Hema stayed at Ritesh's parents' home planning to join her husband once he establishes his career in the United States. It took Ritesh a few months to save money and prepare the rental apartment so his wife could join him in February 2010. Ritesh vividly recalls arriving in Marquette in 2009. He had $500 dollars in his pocket and credits his smooth transition into the community to the people from the hospital who took care of him during the first few weeks. Even a trivial event such as transport from Marquette's airport to Marquette was organized and taken care of. Dr. Tripathi was picked up from the airport by a woman who represented the hospital and he was amazed with how welcoming and helpful she was. She also took him to a local bank so he could open a bank account. She helped him find appropriate winter shoes at a store located in downtown Marquette, and she introduced him to a national chain department store where he was able to find a good jacket that would allow him to remain comfortable during harsh UP winters *(please see* Image 4.1*)*. She even gave him a TV and a few drinking glasses—the first few items in his rental apartment.

Image 4.1 Lake Superior. (Photography by Lali Khalid)

Dr. Tripathi says that the community was wonderful and very welcoming. He is thankful for the assistance he received in the first weeks and months from everyone he met. He said that he fully trusted and relied on the people who he worked with.

When asked about challenges related to settling in Marquette, Dr. Tripathi shared how it is difficult to begin life in Marquette and likely many other places in the United States without a personal car. Due to the scarce public transportation in American cities as well as credit-history requirements at car dealerships, it was very difficult for Ritesh to purchase his first car. His only path to purchasing a vehicle required him to save money for a loan downpayment but also to establish a sufficient credit history. For the first 3 months he took a bus to work, and when the bus was not available, he walked to work. Once Ritesh was ready to buy a car, his experience at a local Marquette car dealership became another milestone in his family's attachment to Marquette. When he and his wife Hema went to a car dealership in Marquette to purchase their first car, they generally felt that everyone at the dealership was very helpful and was trying to recommend the best car within the family's budget that would also allow them to drive safely in heavy UP winters. Once the car was selected, Dr. Tripathi had to secure financing and navigate the tricky process of applying for and decoding details of a car loan. The person who was helping them at the dealership seemed truly invested in the best outcome for Ritesh and Hema. She was extremely helpful and empathetic while trying to navigate securing a loan for a professional immigrant. The Tripathis cannot imagine how different their experience would have been if they worked with a different individual. Dr. Tripathi said that the dealership representative seemed to truly fight for them and was personally surprised and upset with the formal barriers they were experiencing. With her help, the Tripathi's finally received a car loan and purchased a Subaru Impreza.

Home Away from Home: Importance of Culture, Tradition, and Religion

Ritesh and Hema purchased their first home in 2014. The details of their experiences with finding and purchasing a home are detailed in Hema's story in Chap. 3. In his interview, similarly to Hema, he indicated that the key elements of creating his feeling of belonging in Marquette involved housing, children, and activities. Ritesh's and Hema's children were born in Marquette and through his children Ritesh believes he has the opportunity to experience many traditionally Yooper activities. As a family they cross-country ski, they hike in the woods, and they mountain bike. Embracing local culture is equally important to Ritesh as maintaining a connection to Indian culture. As a family the Tripathis travel to attend Indian festivals in Chicago and in Grand Rapids, MI. They visit ethnic Indian communities to provide tangible ways for their children to experience Indian culture, to experience the food, and to participate in a variety of Indian cultural events that are unavailable in the Upper Peninsula. The Tripathis also organize dinners and events at their home in Marquette when they invite their American friends to join them in celebrating their Indian culture. For example, they celebrate the Diwali Festival by inviting 30–40 people to their home. Ritesh puts out lights and Hema cooks traditional food. It is a way for the family to educate their children about the joy of celebrating with the community, and even though they do not have their family in the area, interacting with so many friends and community supporters fills that void.

Dr. Tripathi stressed however that while he exposes his children to Indian culture and the Hindi language, he does not expect them to speak it. For him it is important that his children are part of the community they live in. He demonstrates his commitment to the local culture by various personal contributions including being a coach for a local mountain biking and cycling nonprofit, or by taking the family to volunteer during trash collection at local mountain bike trails. He believes that adopting local traditions is important and it shows respect for the community. Dr. Tripathi shares that Marquette, Marquette is his home, and he wants to give back to the community. "Migration is a two-way street. We adapt to the place we live in and we take responsibility. I am not going to complain that I do not have a temple in Marquette. I knew I was moving to a rural area of the United States. I will travel to bigger urban areas to participate in Indian festivals, but if I am here I will do my best to give as much as I take."

Chapter 5
Claudia

Introduction

Claudia Mankee is an executive administrative specialist at the Office of the Provost and Vice President for Academic Affairs at Northern Michigan University. Claudia was born in Lima, Peru, in the 1960s. Claudia graduated from college with a technical degree in executive bilingual administrative assistant studies and worked in her profession most of her professional life. She moved to the United States in 1993. She is married with two adult children.

Early Childhood and Influence of Place

Claudia was born in a middle-class family in La Punta, Peru. She grew up in a nice house close to the ocean. The environment of Claudia's upbringing created in her a lasting appreciation for water, the sounds of nature, and the beauty of what mother earth provides. Claudia's father was a petroleum engineer and held an MBA degree. Her mother stayed at home and cared for Claudia and her brother and sister. As Claudia recounted her upbringing she indicated that her mother came from a prominent family, and her grandparents were immigrants from Italy. Her family was not wealthy but highly educated and a lot of emphasis was given to solid education, familiarity with the world, and cultural awareness. As Claudia was describing that background I realized that these sentiments of class, aspirations, and family background do not easily translate into US social and class dynamics. For Claudia, emphasizing the Italian roots of her grandparents was intended to highlight that there is a close connection between Claudia's upbringing and a tradition of travel, world exploration, and courageous decisions. In the United States, class perceptions and economic status are often directly proportional. Educational attainment often

W. A. Kusek, *Moving UP North*, https://doi.org/10.1007/978-3-031-95701-7_5

follows money, and money pursues high status and high levels of educational attainment. That is not or has historically not always been the case in Europe and other parts of the world. The way people speak, carry themselves, and the behavioral standards they impose on their families required some resources but was more directly correlated to family history and effort rather than wealth. Pursuing those ambitions for their children, Claudia's parents ensured she attended a bilingual nursery school and her parents arranged for a private English tutor for her up until fourth grade. Claudia wanted to travel the world and explore what other cultures have to offer. That hunger and curiosity about the world is something that is a critical quality Claudia also wanted to instill in her children.

The second cultural quality Claudia acquired in Peru was the need for strong and dynamic social networks. Claudia grew up surrounded by a big community who loved to celebrate holidays. There was social interaction among community members where the ties between neighbors were strong. Food was cooked from scratch and enjoyed with family and friends. Big celebrations were frequent and many holidays were celebrated together.

Claudia's story of migration was sparked by the political environment in Peru. When Claudia became an adult, Peru itself caught in a decades-long internal conflict for political power. The military government coup, active terrorism at its height, human rights violations, and forced sterilization of women were a real risk for people in the country.

For these reasons as well as due to their own objectives, the US military had a presence in Peru. A US Navy Base located in Callao Port was located near the town where Claudia grew up. One of these days, a group of US Navy officers bored with their Officers Club decided to take a walk in a nearby town. It is due to that extracurricular recreational activity that Claudia met her future husband Bill. They walked around town, talked, and exchanged mailing addresses. This was a time when the Internet was just taking off and most communication happened over regular mail. Claudia did not hear from Bill for 6 months. When the letter from Bill arrived, she was surprised and happy, and the couple became pen pals for the next 2 years.

Migration to the United States

In 1991, Claudia's father decided to emigrate from Peru. He remarried, started a business, and was threatened by a terrorist guerrilla group. Due to the risks to his person, Claudia's father moved to Canada. Claudia decided to join him and moved to Toronto at the age of 21. She lived in Canada still corresponding with Bill. The correspondence turned into a real relationship, and Claudia and Bill married in 1993. Once married to Bill, Claudia moved to the Upper Peninsula of Michigan.

Claudia's move to Marquette was not only an opportunity for Claudia to familiarize herself with the city and relatively remote region, but it was also Claudia's introduction to the United States. For the first couple of years, as most immigrants, Claudia was learning about the United States, about the culture, and about her husband's family traditions. Compared to where she grew up, a lot was different in the Upper Peninsula. At the time there were very few immigrants in town and people seemed to have little exposure to other parts of the world. The community was very uniform and traditions were very different from what Claudia knew from Peru.

Compared to her upbringing in Peru and to what Claudia knew from Toronto, her everyday reality in Marquette was very different. Marquette seemed to be very geographically and culturally isolated. She noticed different traditions and interactions with people around her. The weather was harsh and it took her some time to adjust to the rugged, snowy northern climate. That said, as she admits, eventually Claudia learned to love winters.

Culturally, the UP had friendly people but they did not know much about other parts of the world. A culture of individualism was strong and relationships between people seemed loose and weak compared to the level of interconnectedness and interdependence she knew from Peru. Additionally, her everyday life was very different too. Claudia grew up with a stay-at-home mother, sister, father, a nanny, and a very tight community around her, yet in Marquette her only family was her husband, his siblings, and her mother-in-law. There was no extended family, no community to support her, and an admirable but requiring an adjustment focus on local life in the UP. It was that contrast to her own upbringing that drove Claudia to a decision that her children will be educated about the world and exposed to various locations, they will be familiar with different cultures, and they will know Peruvian traditions. She took advantage of the university community and the cultural activities that were happening on campus to expose her children to arts, music, culture, and diversity.

Professional Career

Claudia's story really helped to emphasize how Marquette's growth in its postindustrial reality and Marquette's unique quality blending a small-town atmosphere with bigger-city activities closely relies on the infrastructure and resources provided by Northern Michigan University. In Claudia's case specifically however, Northern Michigan University not only served as a place for intercultural interaction but also became her place of employment. Thanks to the university, Claudia felt she found a path as an immigrant where she could apply her skills and education and translate them into a professional career. As soon as she received her work permit, she began working in the Admissions office where she was employed for 11 years. She also worked in the International Programs office supporting the international community at NMU. When her son was born, Claudia decided to spend more time at home but she wanted to maintain an active employment status. She accepted a 9-month position in the NMU bookstore where she worked for the next 2 years. When her daughter was older, Claudia began working in the Financial Aid office and finally returned to the Admissions office where she worked for another 12 years. When she began considering retirement, a job at the Academic Affairs office opened and Claudia decided to pursue that opportunity because this type of job closely corresponded with her academic degree. In her daily work, Claudia not only uses her education and experience, but due to her upbringing and cultural sensitivity, she serves as the best possible executive-level first contact for the Provost Office.

Home Away from Home and Attitude Towards Migration

Claudia's upbringing and cultural practices from Peru motivated her to stay in touch with her homeland. Today, Claudia still owns property in Peru. She travels to visit her mother and sister in Peru but definitely considers Marquette to be her home. She shared that her hometown in Peru now seems too busy and after two weeks there she is ready to return to Marquette and the calm and scenery of the Upper Peninsula.

While she truly appreciates Marquette today, she recounts that as a mother of young children she was worried about the impact Marquette's remoteness and isolation will have on her children's upbringing. For years she worried that her children were not receiving enough exposure to diversity and other cultures. Claudia felt responsible to incorporate many traditions from her home country in her own home in Marquette. Her children grew up in a home with blended American-Peruvian traditions. Claudia also felt that the community outside of her immediate family would benefit from learning about some of the ways Peru celebrates holidays.

When Claudia moved to Marquette, she was surprised that Americans do not celebrate New Year's Eve in the same way that most people in Peru do. For her first New Year's Eve, her husband took her to a local bar; they had a drink … and went home! This was a culture shock for Claudia because Claudia was used to a big celebration with family and friends. So the following year she invited friends to her home for a New Year's Eve dinner. However, the local culture surprised her again, as everyone left at 10 p.m. In Peru people dance all night, celebrate, and eat lots of delicious food well past midnight.

Claudia was also surprised to learn that many holidays are only celebrated for one day as opposed to Peru where religious holidays are celebrated for a few days in a row and people receive days off work to gather with family and friends.

Claudia missed that community aspect of holidays, so one of the first traditions she introduced to her family and eventually her friends was a big celebration on Christmas Eve. This was a new event in her husband's family. Similarly, to what is done in Peru, Claudia gathers on the evening of December 24th for a celebratory dinner and the family raises a toast with champagne at midnight. Watching TV is not allowed as the family is supposed to interact and celebrate on this special evening. She continues her mother's tradition of exchanging gifts on Christmas Eve but she also respects the tradition of Santa Claus so her children also receive small gifts on Christmas Day. She also adopted the American tradition of Christmas stockings.

Claudia's family also celebrates Easter. She introduced the Peruvian tradition of chocolate eggs with toys but did her best to also provide toy baskets so her children did not feel they stood out among their peers.

Claudia, however, notes that cultural enrichment in her life was not a one-way street. One of the examples she gets excited about is Valentine's Day. When she lived in Peru Valentine's Day was not really celebrated, and in the United States she learned to appreciate the events and decorations and focus on celebrating close and romantic relationships.

Claudia did her best to follow culinary traditions that she knew from Peru. She said it was a culture shock for her to see how much canned foods Americans ate. She tried to cook from scratch as much as she could. Today, the only thing she misses in Marquette is the opportunity to visit a Peruvian restaurant but she understands that she lives in a small city and certain opportunities are simply not available.

When Claudia thinks of the defining qualities of Marquette, she often thinks of very positive things. She loves the lake, spending time on the beaches, and the sounds of water that, despite being much colder than in Peru, bring her comfort (*please see* Image 5.1). She enjoys riding a bike, taking walks, hiking, camping, and spending time at her cabin located on an inland lake.

Claudia truly believes that, despite many challenges including the recent sharp increase in cost of living, Marquette is a wonderful place to raise a family. She shared that she loved taking her children sledding when they were young. She took advantage of the public library and the events that were available to her children in the community. She appreciates the experiences that the university provided for her

Image 5.1 Lake Superior. (Photography by Lali Khalid)

children including the College for Kids Program, performances at the theater, orchestra concerts, and the vast variety of sport events. Her family participated in everything they could and her children turned out to be well-rounded, happy, successful, and worldly human beings. As she built relationships in the community, Claudia learned to really appreciate the small-town environment with friendly people. She appreciates the clean air. She embraced winter and loves how calm and pretty the landscape is when Marquette is covered in snow. Claudia loves thunderstorms and watching the Northern Lights. Ironically, today, it is her husband who asks from time to time if for retirement she would like to spend more time in Peru, and Claudia proudly responds that it is Marquette that is her home. She also adds that "if you roll up your sleeves and work hard, you can be successful here".

I believe that Claudia's ongoing efforts to blend Peruvian cultural practices with her new life in Marquette encapsulate the complex reality faced by many immigrants. In striving to educate her children about their roots while embracing local customs, she mirrors a broader narrative of adaptation—a narrative shared by the other interviewees throughout this work as well.

Chapter 6
Gana

Introduction

Dr. Jelili "Gana" Adebiyi is an assistant professor of Sustainable Food Systems at the Earth, Environmental, and Geographical Sciences Department at Northern Michigan University. He was born and raised in Nigeria. He has lived in Marquette MI for 2 years but his American journey began in 2007 when he arrived in Iowa to attend Iowa State University. He is a professor of sustainability and food systems focused on community planning and development.

Early Childhood and Influence of Place

Dr. Adebiyi was born in a military family in Nigeria. His father's professional career, life dedicated to service, and commitment to his community not only allowed Jelili to attend prestigious schools but it also shaped his personal and professional identity.

Growing up in an environment with significant economic differences among the population as well as poverty and food insecurity influenced Jelili's goals and objectives. Dr. Adebiyi mentioned that he remembers developing fairly early on a strong conviction that food should not be a luxury and there should be enough food to not only feed one's own family but also to be able to produce enough food to sell to others. This philosophy led him to enroll in an Agricultural Engineering major at a local university. While studying mechanical engineering, he got involved in a wide variety of service activities and held leadership roles at his institution observing his country struggling with food insecurity, wars, and challenges related to development. The reality led him to pursue a master's degree in History of Civilization. After completing his bachelor's degree, he moved to Malaysia to search for his

W. A. Kusek, *Moving UP North*, https://doi.org/10.1007/978-3-031-95701-7_6

purpose and meaning and also to further research the importance of agriculture in civilization. This focus on agriculture and learning how food is produced in various contexts and realities is what Dr. Adebiyi cites as the driving force for his migratory journey. Jelili moved to the United States as he wanted to receive the best, applied education.

While his family was in Malaysia, Jelili's wife completed her PhD in Biology.

Migration to the United States

Dr. Adebiyi moved to the United States in 2007 to pursue a second master's degree. He enrolled at Iowa State University to study Sustainable Agriculture. He also completed a master's degree in Community and Regional Planning. He learned about system thinking and how it can be applied in food systems studies which gave him even more motivation to deepen his understanding of place, food, and development. He was ready to move to the Netherlands but instead he decided to move to Michigan to pursue a doctoral degree in sustainability and environmental science from Michigan State University. He explored topics such as policy, ecological foods system, justice, gender, and environmental change, and international development to persistently pursue a deeper understanding of food and place. He also secured a fellowship that allowed him to travel to places such as Malawi, Nigeria, and Mali where he worked on a variety of research projects. All this work was done with the intention to return to Nigeria and he never intended to stay in the United States or in any other country. As it often occurs, however, life took unexpected turns. He secured a postdoc for 2 years at MSU and he served as an assistant professor on a limited term. During that time, he worked on a project in the Upper Peninsula of Michigan as a part of the MSU team. This work was interesting on many different levels but what caught his interest was a comment that his advisor made: "the UP has the same issues related to food insecurity as developing countries." He got inspired to learn more. Dr. Adebiyi quickly learned that issues such as food security, food deserts, and limited access to education in remote areas are all similar issues to what he experienced in the Global South. Not a long time after this project was completed he saw a job opportunity at Northern Michigan University.

His colleagues at Michigan State warned him: "You will not like the weather. It is a different culture up North, you may be discriminated against."

Dr. Adebiyi recalls that despite these warnings, he thought to himself, "Wherever I go I can make it work for me and for my family," and he decided to relocate his family to the Upper Peninsula.

Relocation to Marquette

Dr. Adebiyi and his family moved to Marquette in summer 2022 and very soon Jelili was involved professionally in projects that he was passionate about. His children enrolled in local schools. It was a beautiful summer in Marquette.

The first major challenge the family had to face was the cost of living that was much higher than it was in Lansing and thus higher than he had expected or budgeted for. Rent was not only more expensive but it was very hard to find a place near campus. Dr. Adebiyi was expecting Marquette to be a less expensive place to live but he soon found out Marquette's real estate market was difficult to navigate. Anything from groceries, to rent, to a car mechanic turned out to be much more expensive than it was in lower Michigan. Things took a turn for the worse when his landlord increased rent by $1000 per month and the family was in panic mode trying to find more affordable housing. Dr. Adebiyi was close to giving up and was seriously considering moving with his family downstate. At that moment, the true spirit of the Marquette community came through. One of the neighbors offered to sell Jelili a home that was officially not yet on the market. The neighbor learned that they were looking for a home, that they were from Nigeria, and that they relocated to Marquette for professional purposes. This allowed Dr. Adebiyi and his family to secure a permanent and safe home in Marquette.

While residence was no longer an issue, Dr. Adebiyi had to work through another common challenge for many immigrants, including professional ones—immigration requirements and permits. Even with most supportive employers and communities, immigration paperwork and permits can be daunting. The university committed assistance with immigration but only supported work permits for Dr. Adebiyi, not his PhD-educated wife. Feeling the pressure but also responsibility to support his wife's career ambitions, Dr. Adebiyi decided to pay out-of-pocket for his own immigration attorney to accelerate the processing of his permit and draw a faster path to legal work permissions for his wife. In his account of his journey, Dr. Adebiyi strongly emphasized the challenges of professional migration among two professionally educated and aspiring spouses. Dr. Adebiyi is fully supportive of his wife's career, but especially early in the immigration process one of the spouses often find themselves sidetracked from a career perspective, especially if the other spouse finds professional employment in a rural area. For university employees, sometimes those issues get resolved via programs providing spousal career opportunities. NMU does not offer spousal hire positions so at the moment Fatimat works as an adjunct professor. Dr. Adebiyi says that his wife has a doctoral degree and they both believe there is a bigger purpose behind achieving that level of education. At this point they are still trying to identify the right path for her to deploy her skills.

Home Away from Home: Importance of Culture, Tradition, and Religion

Dr. Adebiyi feels Marquette, MI, is a great place to live for his family. While his daughters are Black and wear hijabs in public, he feels his daughters built strong friendships and feels Marquette is a safe place for his children. An ambitious academic, he is also happy with the quality of Marquette's school system. That said, Jelili recognized the challenges his family had to face with multiple community transitions. He is proud of his children's resilience and flexibility to adapt to a new reality. A lot changed for his children when they moved to Marquette. They left behind many friends from the Lansing area, and his son was involved in an organized soccer club—something he has not yet recreated in Marquette. The family was surrounded by other international and Muslim children so it was easier for them to feel naturally part of a community. When they lived in Lansing, the Muslim and African community was big, organized, and involved. His children had friends, there were community potlucks, and they had access to ethnic food and to religious services.

Despite not having the same opportunities for celebrating their ethnic and religious culture in Marquette, Dr. Adebiyi believes that learning skills necessary to feel comfortable anywhere is essential to pursuing a truly global life. One way they bridge cultural gaps as they change communities is by intensifying the effort and work they dedicate to celebrating their culture at home. As a family, they speak two languages at home. His wife organizes classes for their children. Dr. Adebiyi teaches his children about the culture and traditions of Nigeria, and together the family makes Nigerian food at home. They purchase the ingredients online and often order them from stores in Chicago. While food and cultural education are easier to recreate at home, one larger challenge that remains for the Adebiyi family is access to religious practices. Dr. Adebiyi wants his children to have a good understanding and appreciation of Islam. For Dr. Adebiyi religion is the best way to learn values and structure. Marquette does not have a mosque so the small Muslim community uses the meditation room at NMU's Northern Center to pray, eat, and celebrate religious holidays and traditions. Jelili hopes that the Muslim community in Marquette will grow eventually enough to support a full religious practice. He would like to see a mosque built in Marquette but currently the Muslim population in the area is too small to sustain one. He is ready to serve as a religious leader if an opportunity arises.

Attitude Towards Migration

For Dr. Adebiyi migration is not one event but rather a continuous journey, both figuratively and literally. While he enjoys Marquette, he believes he may eventually move from here as well. The pull factor to arrive in the United States was never purely economic and the move was never meant to be permanent. Instead, Jelili came to the United States to gain knowledge and professional experience, specialized skills, and social capital that he is planning to use to better his home country one day. He plans to move back and help his country develop, thrive, and sustain food production. Dr. Adebiyi emphasized that migration of an individual should have a bigger purpose and the purpose is to make his home country better with everything he learned during his professional journey in the United States. He is very focused on contributing to his community. Dr. Adebiyi is also aware of the brain drain occurring in many African countries and does not want to contribute to it. For him the rules of migration are simple: do not abandon your country; go back and contribute. He would tell students from his country to keep the same attitude. The United States offers wonderful opportunities but use the opportunities wisely and return home to make it a better place for yourself, your family, and your community. That said, he recognizes that as he has experienced before, life can take unexpected turns. Dr. Adebiyi's commitment to sustainable food systems underscores the broader theme of how global challenges—such as food insecurity and environmental degradation—resonate on a local scale in the Upper Peninsula. It is that connection—global to local—that makes migratory journeys so impactful to the rural United States.

Chapter 7
Miguel

Introduction

Miguel Valenzuela is a metallurgical engineer who works for a mining company operating a mine in close proximity to Marquette, MI. He is married to Kendra and they have three daughters. Miguel moved to Marquette, MI, in 2012.

Early Childhood and Young Adult: Influence of Place

Miguel was born in Chile. He was a strong student, with a particular talent for mathematics, geology, and science. He quickly realized that his future would lie in engineering, but he wanted to pursue a more original route than many of his friends who elected to study mechanical engineering. After doing some research he decided to study metallurgical engineering at the University of Concepciόn in Concepciόn, Chile. While a career in metallurgical engineering was promising from an income perspective, he knew it was associated with working in the mining sector where most of the work is in small, remote towns in Chile.

When Miguel was finishing up his studies, he met his American wife who was working in Chile in 2008. His future wife, Kendra, arrived in Concepciόn to attend a religious event. At this point in her life, Kendra was ready to head home to Wisconsin; she did not have many friends or connections in Chile and was pondering the next move for her life. While in Chile, she received a message related to the religious event from a stranger. The stranger was Miguel. They met, connected, and soon fell in love. After Miguel graduated from university, he moved to northern Chile to work for BHP Billiton Escondida Mine which was the largest copper mine in the world. The job satisfied all of his career ambitions; however, he had to move away from his home and be away from his family and fiancé. Kendra stayed at her

W. A. Kusek, *Moving UP North*, https://doi.org/10.1007/978-3-031-95701-7_7

future in-laws' home and Miguel was visiting her on the weekends. After their wedding, Kendra joined Miguel in the north and they started a family. She was there on her own, with no extended family, no support system, with a very young baby, and a husband who was coming back home only for the weekends. It was not sustainable and they knew they had to make a change to be together, to be closer to their extended family, especially that Kendra's mom who lived in the United States was experiencing health challenges. In 2012, they decided to move to the United States.

Relocation to the United States

What is particularly interesting about Miguel's story is that he moved to the United States for many personal and lifestyle reasons, but not for economic ones as Miguel had a well-paying job in Chile with great career prospects. Throughout the interview Miguel emphasized that his migratory journey was a result of family dynamics and making decisions based on the best long-term needs of their family. Initially, Kendra, Miguel, and young Lucia moved to the United States and stayed at Kendra's parents' home in Wisconsin. Upon moving to Wisconsin, Kendra was able to secure a job first. She worked part-time at a local credit union while Miguel stayed at home taking care of their daughter, completing immigration paperwork, and restarting his career. He was comfortable moving to the United States. He had 4 years of professional experience in the mine in Chile and he was certain that he would be able to secure stable employment in the United States. As Miguel kept searching for work, the family's trajectory further changed as Kendra found out she was pregnant with twins. The family's priority was to be together and to live in close proximity to Kendra's parents. They did not want to repeat their experience from Chile.

Meanwhile, the mine in Marquette was being built. Miguel started networking with the mine early in its development to increase his chance of employment near Wisconsin and Kendra's family. A few months later Miguel secured a position in the mine in Marquette and the family relocated to the Upper Peninsula of Michigan.

Marquette a New Chapter

For Miguel and his half-Chilean, half-American family of 5, Marquette was a natural choice due to its proximity to Kendra's family and a unique and promising job opportunity for him. But the family very quickly realized that Marquette may be more aligned with their family values and aspirations than they initially realized. For Kendra and Miguel, the main qualities they were seeking in a city they wanted to call home included a strong sense of community, a safe and family-focused environment, and an active faith community. Marquette exceeded their expectations with regard to those qualities. At the time, as compared to today, it was much easier for the family to find and purchase a home—an event they recognized as finally growing roots in their ideal location.

Miguel is very vocal about his appreciation of Marquette and the Upper Peninsula. He is thankful for the opportunity to live and work in Marquette. He has a strong sense that Marquette is his and his family's home *(please see* Image 7.1*)*. Furthermore, Miguel sees Marquette as representing many of the qualities that make the United States a special country to live in. Miguel is now a US citizen and is very vocal about opportunities that the United States offers. When asked about his perceptions of the United States, Miguel emphasized several aspects of American life that he truly appreciates, focusing especially on the opportunities that he believes the United States provides: "you have the opportunity if you choose to take it." As a practicing Catholic, Miguel appreciates Marquette and the broader United States for the freedom to practice his religion and for a strong Christian community in the Upper Peninsula of Michigan. Miguel also appreciates the United States and for what he perceives to be an emphasis on creating individual responsibility for one's own family. For Miguel that means personal responsibility to provide for the family financially, to secure their future, to raise children in an environment of faith, and even to be responsible for his family's safety. Miguel emphasized the United States' freedom with respect to gun access as an important element of his ability to secure his family. Miguel said: "in the US you have the right and responsibility to protect your family. In Chile, when someone breaks into your home, you have to hide, call the equivalent of our 911, and pray that the police arrive in time to help you." He said that the United States and especially the Upper Peninsula is very safe

Image 7.1 Lake Superior. (Photography by Lali Khalid)

compared to Chile, but he appreciates the opportunity to actively protect his family if needed. When asked about elements specific to the Upper Peninsula that Miguel appreciates, like many other interviewees, Miguel mentioned opportunities for outdoor recreation and specifically camping. Miguel emphasized that in Chile there are not many campgrounds, and for many residents camping is out of reach financially. Due to that context, he particularly appreciates how budget-friendly camping is around Marquette, and he sees that as a great family-bonding outdoor activity.

During the interview Miguel emphasized several times that the active Catholic community in Marquette really helped the family feel at home in the UP. The Valenzuelas not only use religion as a way to define the values of their own family, but use the Catholic institutions in Marquette as a way to be active members of the broader community. The family are significant supporters of Marquette's Catholic elementary and middle school, the Cathedral, and local nonprofit organizations. The opportunity to send his children to a Catholic school was particularly important for Miguel, as that allowed him to replicate his own education model. As a child, Miguel also attended Catholic schools in Chile. The structured, value-based education is what Miguel credits as allowing him to create a strong work ethic that later translated into his professional success. Upon moving to the United States, he was not sure if he would be able to afford to send his children to a private Catholic school in Marquette but he said it became a priority for the family which became a reality thanks to the encouragement and support from the broader Catholic community in Marquette.

Despite a strong and enthusiastic appreciation for Marquette and American culture, Miguel has not forgotten about Chile and Chilean traditions. When asked about what aspects of his Chilean life he transplanted to Marquette, Miguel focused on several holidays and associated celebrations. For example, Miguel mentioned that as a family they all celebrate Saint Day/Name Day. Due to the fact that most people in Chile are named after Catholic Saints, and the Church assigns particular dates throughout the year to celebrate a particular Saint, the day that corresponds with a person's name becomes a special, personal celebration similar to a birthday. It is a tradition in many Christian countries, especially in Europe and the Americas. Miguel explained that he and his wife treat Saint Name Day similarly to what a birthday is in the United States. They sing to their children on their Saint/Name Day, they celebrate with a special dinner, and they offer small gifts. One of their daughter's name days falls around Marquette's local Blueberry Festival, so the family has a tradition to attend the Blueberry Fest every year to celebrate altogether. Miguel also said they celebrate the Chilean Independence Day by inviting their friends to a big dinner on September 18th. Another holiday in their household is the Celebration of the Lady of Mount Carmel Day, patron of Chile. This holiday is celebrated on July 16th and the family goes to mass, cooks a special dinner, and makes a caramel dessert. Finally, they also celebrate November 1st—All Saints Day—by visiting a local cemetery and praying for the dead. Other Chilean holidays and traditions more closely overlap with American holidays, so the Valenzuela's celebrations of Christmas, Easter, or Halloween are more aligned with what other American families enjoy.

After recounting Chilean traditions that his family practices in Marquette and also inventorying all the qualities of life in the UP and the broader United States that Miguel appreciates, Miguel summarized his immigrant story by indicating that to him, home is really not a place in itself, but rather a space where his family can freely define and practice their happiness. While today Miguel visits Chile for work-related purposes and he tries to take his wife and children to visit his parents, with the increased cost of travel they do not visit Chile as much as they would like. Instead, they bring his parents to Marquette so the girls can build a relationship with his family and for his extended family to all participate in what Marquette has to offer.

Chapter 8
Yan

Introduction

Dr. Yan Ciupak is an associate professor of Sociology in the Sociology and Anthropology Department at Northern Michigan University. She served as the institution's Faculty Director for International Partnerships and Initiatives and founded the Grief Support Specialist program.

Dr. Ciupak is a supporter of holistic well-being. She is a certified Yoga instructor and a meditation practitioner. Dr. Ciupak was born and raised in China. She has lived in the Marquette area for more than a decade but her American journey began in 2003 when she arrived in Buffalo to complete a PhD program at the University at Buffalo (SUNY-Buffalo).

Early Childhood and Influence of Place

Dr. Ciupak was born in a small city in the Shanxi Province in China. Both of her parents were educated and worked at a local college. The family lived in a typical Chinese apartment building complex and they led a simple life without significant material aspirations. Dr. Ciupak's hometown, however, offered a variety of cultural events. Yan fully embraced and took advantage of the events that were provided free of charge. Thus, although Yan's family was not affluent, it was easy for them to engage with those cultural offerings. The family also read a lot of books and Yan's father played classical music at home which sparked Yan's interest in world cultures. Yan was surrounded by people with high social capital but not necessarily with the financial means to explore the world. Her hometown was not diverse but there was an awareness of different cultures, countries, and political systems and an urge to learn how others are living their lives in other parts of the globe.

W. A. Kusek, *Moving UP North*, https://doi.org/10.1007/978-3-031-95701-7_8

The family placed a lot of emphasis on education, well-being, spirituality, and connectivity and this emphasis stayed with Yan throughout her life. Like many children from the 1980s China under the One-Child Policy rule, Yan was the only child in her family. That single-child status changed however when her parents divorced and eventually, later in life, she gained a half-brother.

Yan grew up surrounded by her immediate family, extended family, cousins, and neighbors. There was a strong sense of community and strong ties between the neighbors. The physical space was designed for social interaction with a courtyard where the children played and interacted on a daily basis. Neighbors knew each other, and while Yan's social circle did not achieve the levels of closeness mentioned by other interviewees in this book, Yan labels them as "familiar strangers" who were willing to offer help to each other when needed. The community provided a sense of safety and belonging.

Migration to the United States

Yan arrived in Buffalo, NY, in 2003 to enroll in a doctoral program in Sociology of Education at the University at Buffalo (SUNY-Buffalo). She was impressed not only with the diversity of people around her but also with the diversity of public spaces and the perceptible variety of socio-economic zones. Yan found it surprising to see that on one side of the street there was a nice neighborhood while just across the street or one block away existed another community that was economically depressed and where a police car was always present monitoring the activity of the residents. When asked to list other sources of initial, post-migratory surprise, like many other immigrants Yan included impressions of the American food culture. Yan mentioned that she found surprising the amount of highly processed foods with high sugar content.

Relocation to Marquette

Yan's exposure to the Upper Peninsula of Michigan was initiated for personal reasons. Yan is married to a White Michigander who grew up in the lower peninsula of Michigan but attended the Michigan Technological University in Houghton in Michigan's Upper Peninsula and wanted to move back to the UP. Yan's original plan for her life in the United States was to live in a bigger urban area, preferably in San Francisco or New York City, but she was intrigued by her husband's experiences in college, his fond memories of the UP, and the perspective of giving an unexpected opportunity a chance. After arriving in Marquette, she fell in love with the natural beauty of the region and she met many good people who became her friends *(please see* Image 8.1*)*.

Image 8.1 Lake Superior. (Photography by Lali Khalid)

Today, as a multi-year resident of Marquette, Yan loves the area but still sometimes struggles with Marquette's remoteness. She finds it particularly inconvenient to travel to and from the UP. Thinking of what larger cities have to offer, Yan regrets that Marquette does not have many ethnic food options and she longs for cultural diversity. While Yan wishes for diverse cultural experiences in her everyday life, she also makes it clear that there is more diversity in the UP than where she grew up in China. From her point of view, people who complain about the lack of diverse food options in Marquette often compare the UP to big American cities such as Chicago or NYC which are not the adequate comparables. Yan's hometown was culturally homogenous and any products that were foreign were difficult to access or very expensive. This perspective and framework of comparison helped her to stay very realistic about her expectations for the Upper Peninsula. She never really expected the UP to cater to her dietary needs but instead she made the point to figure out how to make her homelife what she wants it to be: a blend of American and Chinese traditions.

Overall, Yan appreciates living in the Upper Peninsula and experiencing American culture through a Yooper lens. She believes that many Americans are self-disciplined. They follow rules better than people follow them in China. For example, Yan notes that in the United States people cross the street in designated areas and often wait for the "walk" signal while in China street interactions are more chaotic and pedestrians think of street crossing signage as an unnecessary obstacle in their arrival to their next destination. Americans are also polite in their interactions with strangers and they are definitely more patient. She also believes that there is more personal freedom which brings a liberating feeling. She appreciates the freedom to study, the respect that Americans have for different points of view, and the freedom of expression and freedom of speech.

In addition to broadly American qualities, Yan learned to value and enjoy nature in the UP. As a child, she recalls spending time at her grandparents' countryside home when she had the opportunity to climb hills and spend time playing outside. Marquette and the UP region remind her of her childhood's happiest times, and she still enjoys spending time on the beach on Lake Superior or Teal Lake near Negaunee, MI. She loves hiking the scenic Sugarloaf Mountain near Marquette and camping with her family.

Home Away from Home: Importance of Culture, Tradition, and Religion

Dr. Ciupak has spent considerable effort to make her home the best place to blend American and Chinese traditions. Spatially speaking, she converted her basement into a meditation room or "Buddha room" to make it suitable for her holistic well-being approach to life. Spirituality is important to Yan. She places emphasis on mindfulness, connectivity, and collectivity rather than on celebration of individuality that is very common in American society. She takes this approach not only to create a home for her husband and children but she also introduces this idea into her teaching practice. She believes that having space between self and emotion is an approach valuable in her classrooms and also for her immediate community. She is an active and highly regarded Yoga instructor. She highly values education, culture, respect, and collective well-being. This sometimes causes tensions in her homelife as her children grow up surrounded by their American friends and the vision of their Chinese mom on how social interactions should occur sometimes differs from what they learn in the community. She has been told by her children that she is too strict and too uptight, and that she should relax more. Despite these family dynamics Yan chooses to instill her Chinese upbringing in her children and holds them to a high educational and behavioral standard. She makes sure that they are receiving a good education and she expects her children to perform academically at the highest level, but also, she has expectations for their performance in social settings. Compared to what Yan is used to, Americans seem much more casual. For example, she mentioned that swearing is unacceptable in an educated Chinese family and she tried to make sure that her children understand that using any form of inappropriate language is absolutely unacceptable.

Similarly, to other interviewees for this book, one of the key ways that Yan expresses her cultural aspirations and norms is via food. From a culinary perspective she has introduced a fusion of culinary cultures at her home. She blends Chinese, American, Mediterranean, and Indian cuisine traditions, and her diet is mostly plant based. It is not uncommon for her to cook two different dinners: one Asian for herself and one more American for her children who prefer a more typical American diet. It is the constant awareness of her dual identity that is part of her everyday life.

She follows some Chinese traditions that are foreign to her children and husband. For example, when Yan was pregnant with each of her children, her parents arrived in the United States to help her care for the babies. They stayed at Yan's home for the first full month after each delivery. They changed the diapers, cooked for the family, washed the dishes, and helped their daughter care for the babies. Discussing this Chinese tradition, Yan smiles and reflects on the fact that while she cannot imagine having children without this support, she has herself become Americanized now, and even though she still plans to help her own children with newborns, she does not plan to be as involved as her own parents were with her children's births.

Yan recognizes that being a Chinese American Yooper places her in a unique position. Her son and husband hunt but she still has not been able to familiarize herself with this Yooper tradition. She feels conflicted sometimes in her beliefs because as a Buddhist she does not believe in hunting yet as a mother she was proud of her son who hunted his first deer at the age of 13. That one example is one of many where Yan finds herself constantly negotiating her identities as a Chinese immigrant and a mother of American and more specifically Yooper children.

Attitude Towards Migration

Dr. Ciupak's story is one that highlights how immigrants, despite fully agreeing with their own migratory decisions, still face ongoing identity struggles and personal difficulties. Yan is happy in Marquette and in the United States. She appreciates her American family, her friends, colleagues, and students at the university. She believes that her unique perspective on well-being also generally benefits her students and provides access to a unique set of skills and viewpoint that many students do not normally get exposed to in rural America. People in the UP often actually seem interested in her cultural background and ask polite questions. That said, Yan indicates that her migratory experience has not always been easy and all her decisions often lead to long-term consequences. She stresses that today she would likely make the exact same decisions, but that does not reduce their consequences. One example Yan highlighted is the fact that she had to renounce her Chinese citizenship to apply for her American one. That was not due to a rule required by the United States but rather by China—China does not allow Chinese citizens to hold another citizenship in another country. Yan was absolutely convinced she wanted to become an American citizen, but the renouncement of Chinese citizenship was emotionally difficult as Asians have strong ties to their home countries.

She also touched upon being a visible minority in a small town and how this affects perceptions and attitudes of some people. She mentioned that in the past she has been asked about American-Chinese politics and the people who asked her about Chinese politics seemed to imply that Yan had some level of influence over China's policies. She has often been put on the spot in personal and professional settings where she was pushed on her perspective or opinion on the actions of China as a country. She fully recognizes that American-Chinese relations are complex and increasingly difficult, but she hopes that people will quickly realize that she is just a single individual who cannot be a representative voice for a society of over one billion people and cannot influence China's actions. Like other immigrants noted in my interviews, Yan also feels that in many cases immigration makes her a foreigner in both countries. In some cases she is not American enough for Americans, but when she visits China she is no longer Chinese enough for people there. When she visits her family in China she is perceived as an American citizen.

Yan feels that negotiating her American and Chinese identities while also incorporating more local, Yooper traditions and practices is just part of her individual journey. Despite stressful situations and sometimes serving as a target for peoples' need to express opinions about the actions of China in the world, Yan appreciates her migratory journey for the career opportunities she experienced and for personal freedoms and accomplishments. Life with her blended Chinese - American family with a lot of regional connections to the UP, is her own and unique world that makes broader global politics less relevant.

Chapter 9
Wael

Introduction

Dr. Wael Khouli is a physician and entrepreneur from Syria who has lived in the Upper Peninsula for almost 20 years. He first arrived in the UP in 2000, later left for Minneapolis, and returned again to Marquette in 2016.

Early Childhood and Influence of Place

Dr. Khouli was born in Damascus, Syria. He grew up in a middle-class family. His father passed away when Wael was a young child leaving Wael's mother to raise him and his 5 siblings with the help of the community. The family lived in an apartment in the city. To navigate the city, Wael relied on public transportation. He biked. He walked in the city. He was a part of a strong community of friends and neighbors and felt close to the broader city community. Unfortunately, as Wael was stepping into adulthood the political situation in his country deteriorated. As a young adult, Wael realized that he wanted to leave Syria and live in a country where opportunities were available and he longed for freedom as he was living in a country troubled by an oppressive political regime. In high school, Wael learned that in Syria he could obtain a medical degree in just six years and that degree would allow him to relocate to the West. He wanted to escape the Iron Curtain and he knew that a medical degree would be his ticket to the opportunities in the West. In his fourth year of medical school, he put his plan into action and was working towards two goals: finishing his degree and passing exams needed to secure residency in the United States. Dr. Khouli thought that when compared to all other Western countries, the United States represented all the qualities he was seeking: freedom, well-paid job opportunities, the most advanced medical equipment, and the best medical training

W. A. Kusek, *Moving UP North*, https://doi.org/10.1007/978-3-031-95701-7_9

in the world. After earning his medical degree at Damascus University Medical School in Syria, he interviewed for a residency position in the United States.

Migration to the United States

Dr. Khouli was able to secure a residency placement at St. Joseph Mercy Hospital in Ann Arbor. His first exposure to the United States was the Midwest. He spent 3 years in Ann Arbor working on his internal medicine residency. He truly enjoyed the city of Ann Arbor. It was a diverse city hosting a sizable community of international professionals, where diverse foods and cultures were part of his everyday life. He had access to Middle Eastern food and many of his colleagues and immediate friends were international. Among his colleagues from his first rotation was a female doctor who grew up in the Upper Peninsula and this is the first time that Wael learned about the "UP."

At the time, Wael did not realize how his legal status in the United States would put him on a path to call the "UP" home. Wael arrived in the United States by receiving J-1 visa status to complete his residency. The J-1 visa supports cultural and educational exchanges and requires J-1 visa holders to return to their country after the completion of their program. One of the ways that individuals can receive an exception and remain in the United States as physicians is to secure employment in a rural area or one that is likely underserved from a medical perspective. Learning of this possibility, Wael reconnected with his colleague from his first rotation and he learned that the Upper Peninsula qualifies as a location that is underserved from a medical service perspective, and if he could work there for some time he would be eligible to obtain his permanent residency in the United States.

Initial Relocation to Marquette

After completing his internal medicine residency, Dr. Khouli arrived in the UP. He provided primary care for patients at the Gwinn and Sawyer Medical Centers but he lived in Marquette. While his introduction to Marquette and the Upper Peninsula was a result of immigration requirements, Marquette and the Upper Peninsula made a very positive first impression.

He was really surprised with the hospital in Marquette, the modern facilities, and the residency program. He was welcomed by colleagues, the community, and his patients. Speaking with a foreign accent posed some challenges, some patients were surprised or at first seemed distracted when he was engaging with them as a provider, but Wael always felt his patients expressed sincere gratitude for his work and welcomed him in the community. He fondly recalls how some patients asked if he was from LA or from India. He understood that the questions were a result of limited exposure to diversity, rather than criticism or judgement. He had friends who worked at the hospital and at the university. Some of them were international but it was a very small international community when compared to his past experience in Ann Arbor.

Wael emphasizes that his UP experience was forever altered and enhanced when he met his future wife during his first year in Marquette. He enjoyed living in Marquette. He appreciated the natural beauty of the Upper Peninsula but initially he felt isolated from busier cities elsewhere. Despite enjoying the scenery, he was not planning to stay here forever. Marquette was friendly and safe but still a small, remote town that lacked diversity of food, people, and culture. Meeting his wife completely changed Wael's perceptions and plans. Michele taught Wael to ski, introduced him to mountain biking, and shared with Wael her love for running. She taught him about pasties, Yooper flannels, and how to enjoy nature. These experiences allowed Wael to not just observe but tangibly interact with Marquette and the environment, forming a connection with the location that ultimately altered his trajectory.

Interestingly, Wael started dating Michele during the fall of the same year that 9/11 events occurred. Dr. Khouli remembers working at the hospital when the events of 9/11 unfolded. As a person from the Middle East he was worried his relationships with his patients would be impacted by backlash towards Arab-Americans. Similarly, he was worried his relationship with Michele could be at risk if her traditional Yooper family became concerned she was dating an Arab man. Wael quickly experienced that his concerns were unfounded and he continued to feel a strong connection to the community. Similarly, not too long after meeting Michele's family Wael felt he earned friendship from them as well. After 4 years in Marquette, both Michele and Wael developed a desire to move to and live in a big city so when a professional opportunity became available, Wael and Michele moved to Minneapolis.

Time in Minneapolis

Dr. Khouli worked at a big hospital system in Minneapolis while his wife worked for the University of Minnesota. Wael was also working on an MBA degree from the Yale School of Management in New Haven, Conn. He flew to Connecticut every other weekend to attend Yale University. The couple lived in the suburbs of Minneapolis. During our conversation Wael provided unfiltered insight into his perceptions of living in the suburbs. Wael indicated he never liked suburbs. He did not appreciate the dependency on cars and a lack of sense of true, dynamic, and organic community. Wael always enjoyed participating in community events and interacting with people representing the complex city population. In the suburbs he felt surroundings and interactions were engineered, generic. In addition to challenges associated with life in the suburbs Wael also realized he missed the trails in Marquette. He disliked that he had to drive for 30 min to access mountain biking trails in Minneapolis.

Thanks to Michele's family connection to Marquette, Wael also maintained a strong and active connection to the UP. The couple visited Marquette twice a year. Every time they were leaving Marquette, they left with a feeling they did not spend enough time with friends and family. The decision to seriously consider relocation was finally motivated by the birth of their first child. When their son was born, Michele realized she really not only wanted to be closer to her family but also simply needed their help. Coincidentally, a new professional opportunity emerged for Wael at the hospital in Marquette. The unique option to meet Michele's goal to be closer to family while advancing Wael's career made moving back to Marquette an obvious choice.

Second Relocation to Marquette

After 11 years in Minneapolis, in 2016 Dr. Khouli was hired as the Chief Medical Officer of the UP Health System Marquette Hospital. The family relocated and rented a house in town. For Wael it is the moment they bought their house that he identifies as the moment he finally felt truly at home in Marquette. The meaning of having his own physical space in the UP made him transition from feeling an immigrant to feeling like Marquette and the UP were his true home. Wael also highlighted how having children who were born in Marquette further increased his connection to the UP. He felt that an integral part of his identity was forever linked to the UP. He also felt Marquette was a great place for his children to grow up. What he appreciates the most about the United States is the freedom to choose and personal freedom in general. In his role as a parent, he hopes to raise his children in a neutral environment which would be in stark contrast to what he did not like about his upbringing in Syria that was heavily directed by Islam. Wael always felt however that he could combine the best of both worlds—celebrate and strengthen his family's connection to Marquette, but at the same time expose them to other cultures. Wael always celebrated his children's Marquette identity, but he also wanted to expose his children to Syrian traditions and more broadly to the world outside of the United States. He truly believes that you can be connected to more than one culture and draw inspiration from the strengths of them both.

Home Away from Home: Importance of Culture, Tradition, and Religion

Wael hopes to be able to one day take his children to Syria so that he could show them the places and locations that defined his childhood. Due to the safety risks of traveling to Syria Wael does not believe a visit there will be possible in the near future. Due to that, Wael invests considerable effort to introduce elements of Middle

Eastern culture at home. He prefers to communicate Middle Eastern traditions through food. Wael cooks meals using his mother's and sister's recipes. His children enjoy traveling to the Lower Peninsula of Michigan to visit Middle Eastern bakeries and restaurants. Whenever the family travels to major urban areas or internationally, they try to visit museums or cultural centers that familiarize their children with Middle Eastern history. Wael taught his kids a few words in Arabic and he reads books to them that teach them about the history and culture of Syria but Dr. Khouli fully understands that his children are growing up in the Upper Peninsula of Michigan and this is their main place of belonging and point of cultural reference (*please see* Image 9.1).

When asked about elements of life in Syria that he misses, similarly to other people featured in this work, Wael focused on the tightly knit, supportive, dynamic communities of people. Wael recognizes that Marquette features a strong sense of community relative to many other locations in the United States, but still does not represent the same level of community interaction and mutual support. While he misses that aspect of life in Syria, Wael also sees his life in phases and embraces the phase of his life defined by Marquette. At this point Marquette is defined by his family, the friendships he developed, and the opportunities it provides—all a reflection of who Wael is today. Wael laughs thinking how a warm summer day represents his dual identity—his backyard hosts views of lake superior, a defining feature of the Upper Peninsula—but at the same time he can smell a freshly grilled kebob.

Image 9.1 Lake Superior. (Photography by Lali Khalid)

Chapter 10
Zhenia

Introduction

Yevheniia Bilan "Zhenia" is a business manager at the Facilities Operations department at Northern Michigan University. She arrived in Marquette at the end of July 2022. Zhenia, her husband Ighor, and their twin children Arkadil and Zlata are refugees. They come from a city of Sumy, Ukraine, which is located 17 miles from the Russian border, and it was one of the first cities to be invaded when Russians attacked Ukraine. Zhenia's migratory journey—sparked by the traumatic disruption of war—provides a contrast to the more gradual professional migrations detailed in earlier chapters. Her story, defined by urgency, resilience, and the quest for safety, highlights the multifaceted nature of migration in today's global landscape. While some migrants plan their journeys, others react and navigate the instinct to survive as well as the natural need to secure a safe and predictable future for their families.

Early Childhood and Influence of Place

Zhenia was born in 1982 in what was then the Soviet Union. Ukraine became independent in 1991 and for the next 33 years her hometown enjoyed rapid development of infrastructure and relative prosperity. Her hometown was always influenced by Russia, Russian culture, and regional politics. The language of instruction at school was Russian, local TV broadcasted in Russian, but despite the many connections with Russia, many people including Zhenia believed that Ukraine could join the European Union and become a modern, democratic nation. It is this very goal that Zhenia worked hard towards in her professional life in Ukraine.

Zhenia graduated from the Sumy State Pedagogical University with a master's degree in English and later she also obtained a master's degree in Business

W. A. Kusek, *Moving UP North*, https://doi.org/10.1007/978-3-031-95701-7_10

Administration from the Sumy State University. She got married, was advancing in her career, and purchased a condo near a beautiful city park near the city center. She worked in the Sumy Chamber of Commerce, one of only 25 organizations like this in Ukraine, and she achieved the position of Vice President. Her achievements, educational attainment, and quality of life placed her among the growing Ukrainian middle class which afforded her opportunities to travel for work and leisure. Professionally, she travelled to many countries in Europe and she visited the United States five times. She visited Chicago, Detroit, Pittsburgh, and a few places in Florida learning how Western societies approach the development of rural regions. She also participated in a Regional Development Internship in Washington, D.C. in 2015.

Transnational Refugee Journey to Marquette, MI

When Russia invaded Ukraine, Zhenia's hometown was blocked by the Russian Army and no food, gas, or any other resources were allowed into the city. When the evacuation corridor was created 2 weeks later, women and children were allowed to leave the city; however, men, including Zhenia's husband, had to stay. Zhenia and her husband decided that she would take their 5-year-old twins and evacuate to nearby Romania. Later with some help from friends in the Netherlands she arrived in the Netherlands where she stayed for 4 months in a rental home hoping that the war would end soon and that she and her children would be able to return home and reunite with Ighor. The end of the war was not coming however. Ighor was also desperately looking for a path to be reunited in safety with his family. Coincidentally, it was the disability of his father that gave him a chance to escape. Thanks to the fact that he was his disabled father's caregiver, he was allowed to leave Ukraine for neighboring Poland under a program to relocate the disabled along with their caregivers. From Poland, Ighor took a flight to Amsterdam where he reunited with Zhenia and his children. The family needed to decide on what to do next. They realized that the war was not going to end any time soon and their children were at an age where they needed to enroll in school. The family had the option to remain in the Netherlands but they were afraid that their children would only learn to speak Dutch and thought that an English-speaking country would provide the children with more opportunities in the future. They came across a message on Facebook. The Crossroads Christian Fellowship of Marquette was offering to sponsor one refugee family and bring them to the UP. Zhenia saw the message as an opportunity to relocate to not only an English-speaking country, but also a country synonymous with opportunity and safety. She quickly got in touch with the church. Although initially the church chose a different family, due to other circumstances the opportunity to move to the United States was eventually presented to Zhenia and her family. The legal foundation for their travel to the United States was based on a specially formed program called "United for Ukraine" that granted Ukrainian refugees the right to temporarily live and work in the United States.

Migration to the United States

The Crossroads Christian Fellowship of Marquette community bought tickets for Zhenia and her family to arrive in Chicago. The community organized basic items for them to begin their new life in the United States. Their welcome package included not only clothes and a residence within the church, but even a car. The family was very thankful because it was very important for them to be together as a family, to be safe, and to have the resources to start piecing back together their life in a foreign country. Zhenia realizes that many Ukrainian children who were not as lucky as her own and were not presented with an opportunity to leave in time had been displaced, kidnapped, or mentally and physically abused. She is thankful for the safe and welcoming environment that the UP offered her children. She absolutely adores the school her children attend. Zehnia was very surprised how accommodating the teachers and students have been. She shared that the school displayed a Ukrainian flag and a music teacher taught all kindergarteners how to sing a song in Ukrainian which brought tears to the family's eyes.

Zhenia mentioned the woman who was the engine behind the process of bringing the Bilans to the UP. Her name was Debrah and she was a senior member of the church community. The Bilans celebrated their first American Christmas with her and her family. Debrah showed them around the UP and she was like a grandmother for Arkadil and Zlata. Unfortunately, Debrah passed away from cancer and Zhenia misses her very much as she was the initial liaison between Zhenia's family and the Marquette community.

The challenges that the family experiences are related to securing stable, long-term permits to stay in the United States. This is partly exacerbated by Zhenia's professional skill set. Zhenia is ambitious, driven, and educated and wants to maximize her opportunities. As of now in 2025, the permit is being renewed on an annual basis, and with the currently changing immigration policies, Zhenia is unsure if her family will be able to stay in the United States or if they will need to look for a new country to immigrate to. Zhenia is also particularly concerned about the short notice period if her status is revoked. This uncertainty causes a lot of stress especially now that Zhenia feels her children have formed a community of friends and feel at home in Marquette.

Zhenia and her family are incredibly grateful for the safety and opportunity afforded to them in the United States. At the same time, on a personal level, Zhenia grapples with reconciling her career achievements in Ukraine, her ambitions, and her current socio-economic standing. Zhenia's college degrees are not recognized, at least locally, to be equivalent to American ones which makes it difficult for Zhenia to apply for and secure well-compensated, professional positions similar to those she had in Ukraine. At the same time, Zhenia realizes that regionally within the UP there are generally limited professional opportunities. Before arriving in the United States, Zhenia and Ighor were planning to quickly integrate into the US job market upon their arrival and secure stable jobs in the modern, booming, American economy. After arriving in Marquette they quickly understood that in order to begin their new life in the United States they have to accept jobs which are available regionally here and now, even if they do not correspond to their experience, educational attainment, or aspirations. They were motivated to work and they both secured employment at a local hotel. Ighor, formerly a small business owner who has a master's degree in Physical Education, started working as a maintenance specialist at the hotel and Zhenia took a job as a sales and event coordinator for the same business. Zhenia arrived in the United States with the understanding that initially she would be willing to work in any position for any job to provide for her family but her professional ambitions are part of her identity and she hopes to be able to progress professionally here at some point. After working at the hotel Zhenia's second surprise was related to navigating employee benefits and insurance. At first the family was not aware that not all jobs in the United States provide benefits so she started searching for employment that would provide her and her family with healthcare and retirement planning options. She applied for a few jobs at Northern Michigan University hoping to secure a position that affords more comprehensive benefits, especially healthcare and retirement, and after a year of failed interviews, she was able to secure her current position.

Home Away from Home: Setting up a New Life in the UP

Life in the United States surprised Zhenia as many things operate differently here, and she is still learning about the United States and her family's new reality. Zhenia said that "visiting the USA is not the same as living in the USA." The limited number of vacation time, expensive healthcare, and differences in the quality of education at school are different from Zhenia's home country. What she really appreciates about Marquette and the UP are the many leisure activities available in this region of the country. She truly enjoys hiking, kayaking, spending time on the beach, and all activities that were not part of her life in the city in Ukraine. Instead of going to coffee shops, visiting shopping malls, and spending time in commercial spaces in Sumy, she spends her time in nature now, with Little Presque being her favorite place to visit *(please see* Image 10.1*)*.

Food is another topic that Zhenia commented on. The cost of healthy, fresh food is high and many Americans do not cook daily. It was surprising for Zhenia to learn that some Americans regularly get food via drive-through fast-food restaurants and they eat in their cars. For Zhenia it is normal to eat breakfast and dinner with the immediate family where all members of the family sit down to discuss the experiences of their day. Zhenia still maintains this tradition. She shops for groceries and cooks meals daily and always sits down with her family to talk and spend time together. Another tradition that the family maintains is related to using regular china and silverware. Zhenia said she was surprised to learn that some families in the United States use paper plates at home. She said that she really misses the culture of celebrating food, serving meals on nice china, and enjoying homemade dishes

Image 10.1 Lake Superior. (Photography by Lali Khalid)

cooked from scratch in an environment which creates an atmosphere of celebrating the time together.

One new tradition that Zhenia formed in the United States is traveling during the winter season. For Zhenia and family, a big celebration is the New Year and that is also a time when the family exchanges gifts. During Christmas however, since Zhenia does not have any other family members here to celebrate with, the family tries to travel to visit other places within the United States. The celebrations that Zhenia focuses on during the year are birthday parties for her children. She noted that her and her husband do not celebrate their birthdays anymore but they put extra effort into making sure that their children have bigger parties. They invite the people they know to join them and they do their best to create special birthday memories for their kids.

Zhenia shared with me that she would like to stay in Marquette. She sees this town as safe and welcoming. If the war ends at home, she would ideally want to move back but she also realizes that her young children are invested in this community and it may be difficult to relocate them even to another part of the United States, not to mention back to Ukraine. She also worries what impact the two weeks of active war has had on her twins. She is not sure if they would ever want to go back to Ukraine even if that option is one day available. As of now she made peace with living in the United States and she hopes to keep receiving opportunities to advance.

Zhenia's experience encapsulates the delicate art of reconstructing a life from forced displacement—a narrative that, although distinct from voluntary professional migrations, similarly underscores the importance of community, cultural continuity, and resilience. What is important to highlight in Zhenia's story is that the unpredictability of her migratory journey may not be over. As this book is being finalized, the legal foundation for Zhenia's presence in the United States is being challenged, placing the family on a path of uncertainty and likely further migration. At the local level in Marquette, there are institutions and people who are connecting Zhenia with immigration attorneys, offering help with planning next steps, and uncovering potential paths of remaining in the United States. But just as those actions are occurring locally, within the family unit Zhenia is facing the emotions and fear associated with needing to secure her next chapter, ideally as far from the ongoing conflict in Ukraine as possible.

Chapter 11
Sonia

Introduction

Dr. Sonia Geschwindt Conolly is a practicing neurosurgeon at the Upper Peninsula Health System Marquette—the main hospital in Marquette, MI. She also serves as a research associate and a clinical director of the Upper Michigan Brain Tumor Center at Northern Michigan University. She was born and raised in Cameroon. She has lived in Marquette, MI, since 2016 but her American journey began in 2000 when she arrived in Connecticut to attend the University of Bridgeport. She received a Bachelor of Science in Health Sciences with a Minor in Mathematics. Dr. Geschwindt Conolly holds a Doctorate in Medicine (MD) from Saint Louis University School of Medicine. Outside of her professional life, Sonia is married with three children. She enjoys living in Marquette and truly believes that this is a perfect place to raise a family. She adds though: "If you choose to be in Marquette, you have to be comfortable to be without your ethnic community."

I was invited to interview Dr. Geschwindt Conolly in her home. When I arrived, I entered her eclectic home. There were a few decorations that indicated Sonia's African heritage but the home overall had a postcard-quality American home aesthetic. The home was full of people: visitors, friends, and family. Her three children were cheerfully talking to their nanny. Her mom, who was visiting from Cameroon, was preparing dinner while chatting with a hair stylist who was getting ready to leave. Sonia offered me a drink and invited me to her office. There was a unique cheerful and warm ambiance resulting from all these interactions and conversations happening all at the same time. That contrast between Sonia's upbringing in Cameroon and the Americanness of her day-to-day life got me excited about the insights I will have been able to collect from this interview.

W. A. Kusek, *Moving UP North*, https://doi.org/10.1007/978-3-031-95701-7_11

Early Childhood and Influence of Place

Dr. Sonia Geschwindt Conolly was born in Cameroon as one of six children. She grew up in an affluent family. Her mother came from a prominent background and her ambitious father became a self-made businessman who was the first in his family to attend college. Sonia grew up surrounded not only by a big family, a broad circle of friends, nannies, and chefs, but thanks to her father's business obligations, she was exposed to guests from various parts of the world who attended dinners hosted by her father. Sonia recalls that many of the guests were male and some were White. While insignificant at the time, Sonia credits that exposure and experience with her future attitude and the social comfort level needed to function in her future profession as a neurosurgeon.

Sonia was a middle child and had to live up to high expectations. She was expected to be perfect, to achieve academically, and to become successful. Young Sonia understood those expectations and she did her best to achieve the goals set forth by her parents. By focusing mainly on academics, Sonia skipped a couple of grades in school and graduated from high school at the age of 16. Students in Cameroon who are from affluent families are expected to study in Germany, Belgium, France, Canada, or generally at Western institutions. Similarly, Sonia's parents had a plan for each of their children's future. Her oldest brother was to become an engineer and to inherit the family business. Sonia's parents planned for her sister to be an accountant and decided that Sonia will be a healthcare professional. Her mother regretted not obtaining a medical degree so she had planned for her middle daughter—the one which seemed so perseverant, smart, and independent—to fulfill her dream. As a result of her family's vision and planning, Sonia attended private Catholic schools as her parents strongly believed that solid education was the key to professional success. Her school was small, offered a lot of individual attention to each student, and was a safe and friendly environment. By the time she finished high school, Sonia had decided she was going to become a heart surgeon. Her parents sent Sonia to Quebec, Canada, to attend college. She stayed with her older sibling and attended classes at a big university in a completely new environment just to learn that Canadian universities have quotas on international applicants to medical schools and they only accept up to 3% of international students. Sonia was 16 and found herself with an uncertain future in a country that was foreign and very different from their home. She was not happy with the cold weather, short days, the big and impersonal classes, and the lack of community around her. Reflecting back on that time in her life, Dr. Geschwindt is wondering if she was simply too young for this challenge of living in a completely different cultural reality, or if she perhaps was experiencing depression and simply did not have the awareness or support system in place at the time to realize what she was experiencing. On top of her daily adaptation struggles she felt that her dream of becoming a heart surgeon greatly challenged. After six months in Canada, she asked her father for permission to return home. To keep his dream and goals for his daughter alive, in Cameroon, her father connected her with a Christian Company that was arranging for Cameroonian students to attend college in the United States. The company helped Sonia enroll in the University of Bridgeport in Connecticut.

Migration to the United States

Sonia arrived in Connecticut to attend a Jesuit university that offered a similar educational environment to what she was used to from home. It offered small classes with individual attention to students. She thrived in this school, took 21 credit hours per semester, and attended summer courses as well. She was a straight A student and enjoyed every minute of her experience. The only thing Sonia recalls as disappointing was the American food and food culture. What she found surprising was the amount of food that people were eating, the all-you-can-eat cafeteria, and the lack of distinct and strong flavors in the food. She was surprised with the amount of food waste at the dining hall and thought of the struggles that many Cameroonians face with food security. Even at her well-to-do home, every piece of meat was accounted for and meals were carefully planned, while in the United States Sonia felt that students lacked awareness of how their eating practices contribute to food waste.

Like many other international students in the United States, Sonia had little awareness or appreciation for the importance of rankings or perceived variation in quality of instruction at various American universities. She did not think about private vs public universities, research vs teaching institutions, or challenges related to obtaining a work visa. She applied for medical schools not realizing that domestic students were being given a priority in admissions. As a high performer academically, who applied to 15 different medical schools, she was only invited to two interviews. The University of St. Louis in Missouri offered her admission under the condition that Sonia's family put 4 years' worth of tuition into an escrow account. Meanwhile, Sonia was also looking for other professional opportunities but she was told that she was overqualified for the positions where she was applying. Faced with a narrow choice for their daughter's future, Sonia's family secured the funds needed to allow Sonia to attend the University of St. Louis where she spent 4 years. She was a hardworking student who always excelled in every class. Dr. Geschwindt credit's her academic success with having both personal and trained self-discipline. She always showed up prepared. She was confident in her abilities. She had a plan but also a lot of expectations to professionally succeed.

For the first two years of medical school, Sonia shared a dorm room with another international student. While that was unusual for medical students, living in a dormitory was another consequence of various adjustments international visitors need to make in the United States. As a foreigner without credit history, Sonia simply could not rent an apartment like her American colleagues were accustomed to doing.

Professional Career

Like many international professional immigrants, Dr. Geschwindt credits her specific career path to a combination of factors that include family's support, academic preparation, personal drive, but also the willingness to take on opportunities when they become available. During medical school, Sonia learned of an opportunity to shadow neurosurgery physicians. She quickly learned that no one wanted to have a neurosurgery rotation as her colleagues perceived neurosurgeons as "mean, not female-friendly" and neurosurgery as a "male-dominated field." For Sonia that perception only increased her interest in this field. Sonia was ready to take up that challenge. She arrived at the first meeting excited and ready to work really hard. She was also not intimidated by perceptions of male dominance, as coming from a traditional African society she was used to existing, proving herself, and even thriving in male-dominated environments. Experiences from her early childhood when she was socially exposed to interacting with her father's business circle was certainly helpful in preparing her for the male-dominated world of neurosurgery. She loved her experience and the team loved her. Sonia was invited to join a research project and her neurosurgery mentors invited Sonia to shadow them even outside of the program in Sonia's free time. Sonia credits her academic preparation but also personal, unrelenting drive for the opportunities she received.

Reflecting back on that time Dr. Geschwindt realizes that she could have come across as a bit too confident if not arrogant at times but she remembers being very comfortable in her skin and in the environment that surrounded her. It may even have been that arrogance that cut through traditional hierarchies that exist in the medical profession in the United States.

One quality that Sonia also stresses is that her most significant or prominent professional mentors were also first-generation immigrants. She admits those people must have understood her experiences and reference points, the struggles that international students face, and the potential that the international community brings to the field. She recalls receiving a lot of support from her mentors. One of them even called her "the son he has never had (sic!)."

As an immigrant from Africa and a woman of color, Sonia's experience in the United States requires her to work through race dynamics unique to the United States. Dr. Geschwindt feels that her upbringing in Cameroon positioned her in a very unique way to work as a surgeon of color, especially in a remote, mostly-White part of the United States. As most international students, she did not know many details or intricacies related to racial relations in the United States, and thus, she never even thought about racial differences in the United States, discrimination, or any of the challenges that are part of the daily experience of life for African Americans. Even today, after more than two decades in the United States, Sonia distinguishes her experiences as an African from the history and culture of African Americans. Sonia identifies as an African and an American, but not as African American. When thinking of her identity after becoming a US citizen, she feels American first and African second: two, separate identities. She has spent all her adult life in the United States. She married an American. Her children are American. And it is in the United States where she created her home and her professional and family life, and it is in the United States where she made many life-lasting friendships with people from all over the world.

It is not that her professional life was all without challenges. At the very same time that 23-year-old Sonia was becoming successful in her neurosurgery world, her father changed his rhetoric. He informed Sonia that he believed it was time for her to return to Cameroon, become a pediatrician, and open a clinic in her hometown. First and foremost, in her fathers' eyes, Sonia should get married and have children and the 7-year neurosurgery residency was not a good idea. He did not believe that neurosurgery was right for women and in fact he did not believe that any man would marry Sonia if she became a neurosurgeon. When he learned that Sonia was going to follow through with her plans and enrolled in neurosurgery residency, her father cut her off from any funding that would allow Sonia to finish her studies and participate in residency interviews. Facing a real challenge to her future, Sonia was able to enroll in a residency program at Thomas Jefferson University in Philadelphia thanks to the financial assistance and emotional support from her professional mentors—those same ones that were initially described as "mean" and "against women in neurosurgery."

Relocation to Marquette

Dr. Geschwindt's migratory decisions and where she established roots for her family were not just a result of careful career planning and execution, but also heavily influenced by personal factors and unexpected life situations. At the time Sonia was ready to enter the job market, she got married and became pregnant with twins. Her pregnancy was challenging, requiring several weeks of bedrest. This difficult time changed all her priorities. Reflecting back, Dr. Geschwindt says that if she had her children earlier in life, it is possible she would have not become a neurosurgeon but at the time she was already on the job market and ready to move to a new place that would accommodate the needs of both her neurosurgery career and her young children. The vulnerability of her babies and the difficult pregnancy made her realize that she wanted to live in a place where she would be able to continue her professional career while also being an involved mother. As such, Sonia was looking for a small community. While she had a successful interview in North Carolina, she felt that race was too much of a factor in her interview process and the group did not appreciate Sonia as much for her skills, but rather really wanted to hire a Black surgeon. She kept looking for the right fit, and when she arrived for her interview in Marquette, she immediately knew she wanted to live in the "UP" *(please see* Image 11.9*)*.

When asked about the initial qualities of Marquette and the Upper Peninsula of Michigan that were so striking, Sonia lists a few things but focuses on one predominant characteristic. Dr. Geschwindt liked Marquette due to the natural beauty of the

Image 11.9 Rocks near Lake Superior. (Photography by Lali Khalid)

place, the small and friendly community, the convenience of a small town. However, most importantly, Sonia felt that people were genuine here. Marquette was real. Her colleagues wanted to hire her for who she was, as a person and professional. She felt respected as a human being and from a professional standpoint she knew that her skill set would be meaningful as rural communities struggle with attracting physicians. It is through these experiences that Dr. Geschwindt's migration narrative also engages with the complexities of race in a predominantly White rural setting, where her identity as a Black, internationally educated physician both challenges and enriches local perceptions of race, and the significance of race in general.

Upon moving to Marquette, Dr. Geschwindt admits that at first, she tried to fit into the broader community aesthetic. She went to Kohls and bought some outfits that she believed would make her look more relatable to her patients and the people she worked with. But very quickly she realized that everyone around her liked her for who she was. They appreciated her big city fashion choices and her braided hair. Her patients loved how she carried herself. Dr. Geschwindt concludes: "as long as you own who you are and you are good at your job, Marquette is a gem."

Home Away from Home and Attitude Towards Migration

Dr. Geschwindt loves living in Marquette. Her children are happy here. The family made many friends. Sonia shares that many of her immediate friends come from international backgrounds. She calls her small circle of friends cosmopolitan as people come from various backgrounds yet respect each other's uniqueness. She emphasizes that the unique experiences that her friends bring to the table make her personal life interesting without judgment or competition. Everyone works hard in their own professions and when the friends hang out they take off their professional hats and enjoy each other as humans. It is a group of friends who love good food, good laughs, and interesting stories. Part of that experience Sonia also credits to Marquette. Marquette and the UP offer many activities that test human qualities over professional achievements or socio-economic status. Sonia cross-country skis with physicians and teachers or city-workers; Sonia enjoys the beach or lakes with hospital administrators and retired military veterans. Sonia participates in fundraising marathons with nurses but also power line workers and students.

When asked about her experience living not just in Marquette but in the United States in general, Sonia reflected on the fact that what appreciates about American culture are the opportunities that are available for women. While Dr. Geschwindt knows that not everyone has equal opportunities, the opportunities in the United States seem infinite when compared to countries like Cameroon. Sonia highlighted that in Cameroon she feels there is a lot of pressure on women to perform specific, often gendered roles, while in the United States she feels empowered to maximize her potential. Dr. Geschwindt says that it was liberating for her to move to Canada and later to the United States. Despite the appreciation for more individual liberty, Sonia also highlighted that she often needs to navigate both cultural standards. If there is anything that she still strongly adheres to from her culture—it is the perception that no matter what professional achievements she has accomplished, she is still a woman who needs to care for her family. When she enters her home, she sets aside her identity as a doctor. She stops managing the situation around her. Even today when I interviewed Sonia, her mother reminded her that she is a wife and she should change into a nicer dress because her husband was soon scheduled to come home from work. We laughed about this comment but we both recognized these conflicting cultural signals and we understood the internal struggle that many international women face even if they live thousands of miles away from their home countries.

Having established a fulfilling life in Marquette, Dr. Geschwindt no longer misses Cameroon and she does not visit her birthplace often. Perhaps it is related to the fact that she left her hometown when she was 16 and she did not visit often since then. Socially, she feels disconnected from her home country. Other than her family, she does not have other social ties to her hometown. Her family visits her in Marquette regularly but she does not feel the urge to go back. She does not believe that her home country is even safe enough at this point to travel there with her children. When asked if she will stay in Marquette forever, Sonia hesitates—as a migrant who lived in three countries, she realizes that life can take unexpected turns. She also loves different cultures and loves to travel. Dr. Geschwindt also admits that perhaps when she retires she will choose to live in different places around the world. But for now, Marquette is her home and a great home indeed.

Chapter 12
Conclusions

Introduction

The stories featured in this book represent individuals from various cultures, genders, ethnic traditions, and personal trajectories. Despite these differences they all currently call Marquette, MI, and maybe even more broadly—the Upper Peninsula of Michigan—home. The objective of this book was to spotlight people that are often overlooked in the broader discourse on migration, professionals who often play essential and always culturally and professionally enriching roles especially in rural communities like the "UP." While each interviewed individual represented a unique story and perspective, I specifically enjoyed finding common themes across those stories.

Key Observations

First, and probably most obviously, it is important to highlight that professional immigrants interviewed for this book moved to Marquette for one of two main reasons: professional opportunities or family-related matters. Those who moved here for family-related reasons could leverage their advanced skill set to more easily secure gainful employment or an interesting career path. When analyzed together, these narratives reveal that while professional immigrants come to the Upper Peninsula for a variety of personal or career-focused reasons, they all engage in the complex act of negotiating dual identities—preserving home-country traditions while embracing new ways of life. Their stories highlight that integration is a dynamic, ongoing process that redefines what it means to "belong."

The uniqueness and complexity of their migratory journeys also lends itself to the fact that international professional immigrants do not initiate chain migration to

W. A. Kusek, *Moving UP North*, https://doi.org/10.1007/978-3-031-95701-7_12

Marquette or the UP. Their personal migration is not followed by subsequent migration of people from the same area, country, or family, and if it is, it is likely coincidental rather than linked to these individuals specifically.

The most important quality represented by these individuals is that they uniquely blend celebrating their traditions and cultures with a clear goal and aspiration to be integral members of Marquette's general community. While these individuals try to expose their children to their home-country language, cultural traditions, celebrations, and culinary experiences and they emphasize their countries and cultures of origin, at the same time they celebrate and appreciate where they are now. These individuals are not really looking to join or create a familiar cultural group; they rather aspire to become an integral member of the broader city or regional community.

For these individuals, becoming an integral or contributing member of the local community starts with deploying their skill sets. Professional immigrants in Marquette are often employed in the fields where they can contribute their skill set to fill a professional need within UP's generally rural community. The people interviewed are doctors, professors, and project managers, community leaders, and community caregivers. And regardless of country of origin, they all represent a high level of fluency in English. Additionally, living in Marquette, they also recognize and appreciate the recreational opportunities available in this area and actively pursue at least some level of familiarity with activities that define "Yooper." For some of the people interviewed for this book, Marquette was their first opportunity to experience camping, mountain biking, water sports, and definitely winter sports like downhill skiing or cross-country skiing. But all interviewed people at least added hiking in the unspoiled wilderness to the list of their favorite activities.

The interviewed immigrants grew up in communities which were organized and active and provided ample opportunities for true community interaction. Due to that background, they particularly appreciate that Marquette is a place where people recognize each other after some time, and locals are open to forming new friendships and connections. Immigrants appreciate the walkability and a sense of place in this town.

It is clear in all the stories included in this book that professional immigrants are fulfilling their "American Dream" either through advancing professionally, achieving materialistic success, enjoying freedom as defined by the cultures they came from, building a strong community and friendships in a new locale, or raising their children to be global citizens. The American Dream may not have to be exactly the same as it is for many Americans who grew up in the United States but the common denominator is advancement in American society.

Despite this appreciation for their opportunities and motivation to form connections and friendships in Marquette, I found that the process of interviewing these individuals reignited the pain of difficult choices. For some of the participants, the process of interviewing for this book project was the first time in their lives that they took the time to stop and reflect on their immigrant experience. As my interviewees recounted their migratory stories, they recognized and voiced concerns about their lives, questioning the choices that they made for their own children and how they changed the path for their families. This experience showed me that immigrants, even those who are professionally accomplished and financially stable, worry about the families that stayed behind in their home countries and they do their best to keep in touch with their loved ones and/or to support them either emotionally or financially. They also sometimes question whether their personal success was worth the weakening of connections to their families and cultures and recognize that a decision to migrate is never a simple or straightforward personal journey.

Lastly, interviewing these professionals refocused my attention on complex themes related to privilege, especially in the context of broader class discussions in the United States. Professional immigrants to the United States can be and often are perceived as privileged. These observations are often based on the fact that a professional migratory journey requires individuals to have high levels of educational attainment and simply resources or at least experiences to make decisions and choices that allow them to pursue professional opportunities in the United States. That said, that perspective fails to account for the frequent and broad spectrum of challenges and obstacles that each of them had to overcome to achieve the positions they are in today and also does not account for the multitude of fast, life-changing decisions that impacted the life trajectory of their families or the frequent if not constant internal debate that these individuals conduct about their past and upcoming choices. The success stories presented in this book are of those who showed perseverance, overcame obstacles, and became successful in their fields and respected in their community. But these stories should also be a reminder of the many professional immigrants who for a variety of professional, personal, or immigration-related reasons had to write the rest of their story outside of the United States.

Taking all these stories, findings, and perspectives together, I am hoping that American readers of this book are left with three main points of reflection.

First, I recognize that in the United States today, when people hear the word "immigration" they have a very specific connotation. I am hoping this work highlights that immigration stories are complex, reasons for migration are complex, and motivations for moving to the United States as well as staying in the United States are also often very personal and not straightforward. Gaining something in the United States often means leaving something else behind.

From a daily life perspective, I hope this book shows how migrants can effectively blend identities—how they celebrate their home cultures and practices, how they recreate a sense of home when away from home, but also how they embrace their new homes, influence their social circles, and contribute to the local cultural and economic reality. I was personally amazed how migrants use food and food culture to both recreate a sense of their home identity and also to build new friendships and a new community by using food as an anchor. I was also particularly impressed how uniformly across most if not all interviews, the number one thing migrants missed from their home countries is a strong, active, and vibrant community. People interviewed in this book are used to celebrating holidays together, relying on each other for community child-rearing, and feeling a strong sense of community bonds; in the United States at least at first, most migrants felt isolated and were surprised by how infrequently or how formally their American spouses, friends, or neighbors interact; as such immigrants to the United States feel the responsibility to introduce stronger community relationships and become the driving force behind more elaborate celebrations of holidays, closer support for new parents, or simply care for one another.

The third outcome of this book, while not intended by the author, seems to underscore that towns like Marquette, MI—smaller towns in the rural Midwest—are not only hosts to immigrants from all over the world, but feature resident populations that in many cases are welcoming to people of all cultures and backgrounds. The oversimplification of the American reality where big cities are welcoming for foreigners, but rural areas conservative and closed to diversity, is simply not the experienced reality of immigrants featured in this book. The interviewees selected for this book were only selected for interviews due to their educational or professional achievement without any knowledge by the author of their perceptions of living in Marquette. Yet, while not all stories and experiences featured in this book are pleasant or devoid of assumptions made on the basis of ethnicity, the overall picture is optimistic. Interviewees representing European, Asian (both Far East Asian as well as Middle-Eastern heritage), African, and South American cultures generally do not feel discriminated and, maybe even on the contrary, feel that rural Michigan has become their true home where they built lifelong friendships and where they appreciate opportunities to contribute to local communities. While not universal across interviews (and particularly not applicable to migrants from Europe who come from democratic traditions), many interviewees also appreciate living in the United States for the ability for them to express their personal freedom and ultimately for the career opportunities that became available to them regardless of their ethnic background or country of origin. It was furthermore interesting to see a pattern among interviewees indicating that immigrants want to be mainly recognized for their educational attainment and professional skill set and want to feel part of their local Michigan community rather than be recognized for their ethnic background or seek out culturally specific groups and events. To be more accurate, the immigrant stories indicate the desire for two objectives: first, for the broader community to understand and most importantly to not oversimplify or misperceive the migrants' home cultures and, second, the desire for the immigrants to be integral and valued community members that are valued for their educational, professional, and cultural contributions and not only due to the uniqueness of their heritage.

While most content for this work was written in 2024, in light of recent shifts in national immigration policy—from heightened enforcement measures to revised visa regulations—I find it important to at least briefly explore and speculate about whether the local connections documented in these interviews may face new challenges. While the narratives in this book vividly capture how professional immigrants forge deep, resilient bonds in small-town environments like Marquette by leveraging personal connections and community-focused local dynamics, it is important to acknowledge that national policy changes can fundamentally reshape these local dynamics. Future policy shifts may alter access to employment opportunities, change community resource allocations, or even shift public attitudes towards immigrants, potentially reconfiguring the supportive networks that are so critical to immigrant integration. Thus, as these local stories of resilience and belonging continue to evolve, I do find it important to stress that they remain intertwined with—and vulnerable to—the broader national debates over immigration policy.

Index

C
Cameroon, 71, 72, 74, 75, 77
Chile, 47–51
China, 8, 53–55, 57, 58, 68
Community, 2–5, 8, 10–12, 15, 16, 18–21, 23, 24, 26–30, 32–34, 36–45, 48–50, 54, 56, 59–61, 63, 67, 69, 71, 72, 74–76, 79–83

H
Home away from home, 2, 18–20, 27–28, 33–34, 38–40, 44, 56–57, 63, 68–70, 77–78

I
Immigrant identity, 18
Immigration, 1–3, 5–11, 19, 20, 43, 48, 58, 60, 67, 70, 81–83
India, 8, 23–27, 29–31, 60

N
Nigeria, 16, 41–44

P
Peru, 35–40
Poland, 3, 12, 15–18, 20, 21, 66
Professional immigrants, 1–3, 5, 7–11, 27, 29, 33, 74, 79–82

R
Refugees, 5, 11, 12, 65, 66

S
Sense of belonging, 5, 9
Sense of place, 2, 3, 9, 80
Syria, 59, 60, 62, 63

U
Ukraine, 65–70
Upper Peninsula of Michigan, 2, 4, 7, 9, 11, 17, 26, 27, 36, 42, 48, 49, 54, 63, 75, 79

W
Well-being, 18, 53, 54, 56, 57

Y
Yooper, 8, 28, 33, 55, 57, 58, 61, 80

W. A. Kusek, *Moving UP North*, https://doi.org/10.1007/978-3-031-95701-7

The manufacturer's authorised representative in the EU is Springer Nature Customer Service Centre GmbH, Europaplatz 3, 69115 Heidelberg, Germany. If you have any concerns regarding our products, please contact ProductSafety@springernature.com

Printed and bound by CPI Group (UK) Ltd, Croydon, CR0 4YY

30/09/2026

02225471-0003